A PRACTICAL GUIDE TO ANALOG BEHAVIORAL MODELING FOR IC SYSTEM DESIGN

A PRACTICAL GUIDE TO ANALOG BEHAVIORAL MODELING FOR IC SYSTEM DESIGN

by

Paul A. Duran

KLUWER ACADEMIC PUBLISHERS
Boston / Dordrecht / London

Distributors for North, Central and South America:
Kluwer Academic Publishers
101 Philip Drive
Assinippi Park
Norwell, Massachusetts 02061 USA
Telephone (781) 871-6600
Fax (781) 871-6528
E-Mail <kluwer@wkap.com>

Distributors for all other countries:
Kluwer Academic Publishers Group
Distribution Centre
Post Office Box 322
3300 AH Dordrecht, THE NETHERLANDS
Telephone 31 78 6392 392
Fax 31 78 6546 474
E-Mail <orderdept@wkap.nl>

 Electronic Services <http://www.wkap.nl>

Library of Congress Cataloging-in-Publication Data

A C.I.P. Catalogue record for this book is available
from the Library of Congress.

Excerpts from Analogy materials are included with the permission of Analogy, Inc.
Such materials may not be copied or reprinted without the express written permission of
Analogy, Inc., 9205 SW Gemini Dr., Beaverton, OR 97008

Copyright © 1998 by Kluwer Academic Publishers

All rights reserved. No part of this publication may be reproduced, stored in a retrieval system
form or by any means, mechanical, photo-copying, recording, or otherwise, without the prior
the publisher, Kluwer Academic Publishers, 101 Philip Drive, Assinippi Park, Norwell, Mass

Printed on acid-free paper.

Printed in the United States of America

To the memory of my uncle Bernard P. Duran -
a loving father to his son Diego,
and a revered Veteran who gave blood for his country in Vietnam.

TABLE OF CONTENTS

1.0 Introduction 1

2.0 Modeling and Simulation Background 5

  2.1 Analog Simulation.. 5
    2.1.1 The SPICE Simulator 6
    2.1.2 AHDL Simulator 7
  2.2 Digital Simulation 8
    2.2.1 The SPICE Simulator 8
    2.2.2 Digital HDL Simulators 9
  2.3 Mixed-Signal Simulation...................................... 9
    2.3.3 Mixed-Mode HDL Simulator 10
    2.3.4 Native Mixed-Signal HDL Simulator...................... 11
  2.4 HDL Basics 12
    2.4.1 Digital HDLs 13
    2.4.2 Analog HDLs (AHDLs) 15
    2.4.3 Mixed-Signal HDLs 18
  2.5 Abstraction... 23
  2.6 Modeling Continuum.. 25
  2.7 Model Precision and Accuracy 28
  2.8 IC Modeling... 30
    2.8.1 SPICE Device Modeling 30
    2.8.2 SPICE Macromodeling.................................... 32
    2.8.3 Analog Behavioral Modeling 33
    2.8.4 Analog Behavioral Macromodeling 34
    2.8.5 Algorithmic Modeling 35

2.9 Applications of Analog Behavioral Modeling....................36
   2.9.1 Dynamic Specification 36
   2.9.2 System Design 37
   2.9.3 IC Design 38
   2.9.4 Virtual Test....................................38
   2.9.5 Other Applications 39

3.0 Methodology..43

3.1 Design Methodologies......................................43
   3.1.1 Bottom-Up Design 44
   3.1.2 Top-Down Design 45
   3.1.3 Hierarchical Design 47
3.2 Analog Behavioral Modeling Methodology 53
   3.2.1 Specification....................................54
   3.2.2 Development....................................55
   3.2.3 Verification.....................................56
   3.2.4 Documentation 56
   3.2.5 Release Control..................................56

4.0 Basic Building Blocks 59

4.1 MAST Mini-Tutorial..59
4.2 Electrical Sources 63
   4.2.1 DC Current Source...............................63
   4.2.2 DC Voltage Source...............................63
   4.2.3 Voltage Controlled Voltage Source 64
   4.2.4 Current Controlled Current Source....................65
   4.2.5 Exponential Sinusoidal Voltage Source 66
4.3 Voltage Arithmetic 69
   4.3.1 Voltage Addition, Subtraction, Multiplication, and Division 69
   4.3.2 Voltage Differentiation and Integration 72
4.4 Electrical Primitives 75
   4.4.1 Resistor..75
   4.4.2 Capacitor.......................................76
   4.4.3 Inductor..77
   4.4.4 Ideal Diode.....................................77
   4.4.5 Ideal Transistor..................................79

Contents

5.0 More Building Blocks 85

5.1 Analog Models 85
    5.1.1 Ideal Transformer 85
    5.1.2 Peak Detector 89
    5.1.3 Sample-and-Hold 92
    5.1.4 Non-inverting Schmitt Trigger 96
    5.1.5 Voltage-to-Frequency Converter 99
    5.1.6 Frequency-to-Voltage Converter 102
5.2 Digital Models 106
    5.2.7 AND Gate 106
    5.2.8 Multiplexer 108
    5.2.9 D-Latch 109
5.3 Mixed Signal Blocks 113
    5.3.1 Voltage Comparator 113
    5.3.2 Pulse-Width Modulator 116
    5.3.3 Analog-to-Digital Converter 120
5.4 Mixed-Signal Interface Models 126
    5.4.1 Analog-to-Digital Interface Models 127
    5.4.2 Digital-to-Analog Interface Models 129
5.5 Mixed-System Models 131
    5.5.3 DC Motor 131

6.0 IC System Examples 137

6.1 Distributed Power Supply 138
    6.1.1 System Overview 138
    6.1.2 Model Implementation and Verification 141
6.2 Automotive Ignition System 154
    6.2.1 System Overview 154
    6.2.2 Model Implementation and Verification 155
6.3 Audio Test System 164
    6.3.1 System Overview 164
    6.3.2 Model Implementation and Verification 165
6.4 Digital Communication System 181
    6.4.1 System Overview 181
    6.4.2 Model Implementation and Verification 184

Appendix A 192

A.1 Buck Averaged Converter Netlist............................192
A.2 Forward Averaged Converter Netlist........................194
A.3 Cascaded Converter Netlist 196
A.4 Top-Level Automotive Ignition Netlist 200
A.5 Electronic Control Unit Model Code 201
A.6 Position Sensor Model Code 202
A.7 Spark Plug Model Code 203
A.8 Top-Level Audio Test System Netlist 205
A.9 Loudspeaker Subsystem Test Netlist 211
A.10 DSP Subsystem Test Netlist...............................212
A.11 Voice Coil Model Code 213
A.12 Wind-Drag Model Code 214
A.13 Successive Approximation Register Model Code 218
A.14 Digital to Z-Domain Converter Model Code 220
A.15 Nonlinear Spring Model Code 221
A.16 Top-level Digital Communications System Netlist 223

Index 227

About the Author 231

Preface

Simulation and modeling of integrated circuits (ICs) play a vital role in today's highly competitive electronic industry, where time-to-market is crucial in succeeding in an industry where technology changes rapidly from one year to the next. Innovations in Electronic Design Automation (EDA), along with the ever increasing computing power of workstation and desk-top computers, have decreased product development cycles and improved design performance. Advanced process technologies enable a variety of analog and digital components to be easily integrated. Very-large-scale and ultra-large-scale integration, better known as VLSI and ULSI, drive the electronic products down in size and up in complexity. ICs are evolving into complete systems

Because of this integration, the production of an IC from specification to design, fabrication, and test is becoming more complex and expensive. Management has been searching for ways to make this process as efficient as possible through design automation, process refinement, etc. Yet with all the software and computing power, the goal of getting it right the first time has not been consistently achieved. There is no exact answer to this problem. Simulation and modeling is not the solution in its entirety, but it has proven to be an integral part of all design and product development processes. And when approached methodically, has proven to be the key to saving time and money.

The most successful designs can be attributed to a good proven design methodology. However, because of the increasing complexity of IC design, those methodologies need to be revisited to encompass the challenges of large mixed-signal IC systems. New simulation technologies have evolved to address the issue of increasing design complexity. Embracing these new technologies will become an essential component of adapting current methodologies to meet the requirements of advancements in design integration.

One such advancement in simulation technology, is the development of hardware description languages (HDLs) and behavioral simulators. Digital HDLs have evolved to address the challenges of large digital designs. Similarly, for large mixed-signal designs, analog HDLs (AHDLs) have evolved. Analog behavioral modeling (ABM) can be used to decrease time-to-market while improving the overall design quality of large IC systems. It is never too soon to begin to understand how to improve existing design methodologies to be successful in the very competitive industry of semiconductor systems.

Acknowledgments

I would like to gratefully thank the following individuals for their assistance in editing this text: Tom Cummings, Ira Miller, Marty Brown, and Mike Vicker.

I would like to thank Lee Tang for the design of the cover art work.

I wish to express my gratitude to those at Analogy Inc., who assisted with various aspects in creating this text: Ian Getreu, Ron Evans, Doug Johnson, Chris Brigden, Steven Mayes, and Mike Donnelly.

I gratefully acknowledge the general support and motivation from the following individuals: Siew Ling Ng, Mark Lee, Tina Tsai, Helen Chang, Kiet M.Van, Lynn Ford, and Bob Bublitz. I apologize to anyone that I may have neglected to mention.

Finally, I would like to thank my parents Paul and Marina, and my siblings Juanita and Lawrence for their love and encouragement.

Author's Note

The author has made every effort possible to ensure the accuracy of the material in this book and examples provided therein. The author does not take responsibility for using this material in design. The material in this book is intended for academic training purposes only. The author and publisher make no warranty of any kind, expressed or implied, with regard to the material presented in this book. The author shall not be liable for damage in the connection with, or arising out of, the furnishing, performance, or use of the material in this book.

It is the responsibility of the reader to test all models presented in this book. The reader should develop a test procedure to allow for the complete understanding of the applications, prerequisites, and uses of any model developed from the material presented in this book.

1

Introduction

Today, IC systems are becoming so complex that the circuit designer needs to take a systems approach to design in order to assure that the individual sub-systems interact with each other as expected. It is also important to verify that the IC is compliant and functional with the respective application.

The ultimate goal of this book is to demonstrate a practical and relatively inexpensive approach to decreasing the time-to-market while improving the quality of IC systems. The key is verification through simulation. There is no easy solution to the problem of simulating entire systems, but the EDA industry is moving closer to a better and more robust solution with the evolution of analog extensions to Verilog and VHDL - these languages will be covered in more detail in Chapter 2. This book will focus on analog behavioral modeling and simulation using analog hardware description languages (AHDL's). Specifically, it will elaborate on existing IC design methodologies and comment on how analog behavioral modeling can be adapted to these methodologies for the purpose of saving time and money in the design of IC systems.

This book will present a methodology to abstract an IC system so that the designer can gain a macroscopic view of how sub-systems interact, as well as verify system functionality in the respective applications before committing to a design. This will prevent problems that may be caused late in the design-cycle by incompatibilities between the individual blocks that comprise the overall system, or incompatibilities between the IC and the respective application. We will focus on the techniques of modeling IC systems through analog behavioral modeling and simulation. We will investigate a practical approach by which designers can put together these systems to analyze topological and architectural issues to optimize system performance.

This book is intended for the practicing engineer who would like an introduction to and gain practical knowledge in applications of analog behavioral modeling. The contents of the book is designed for the amateur modeler. Even though the book is geared towards IC design, the concepts presented can be applied to many engineering disciplines. It is assumed that the reader has some experience with IC design. A knowledge of Verilog, VHDL, or MAST® [1], will also be useful for the reader. However, it is not necessary to be fluent in any hardware description language, since the information presented is focused on model implementation as opposed to the actual implementation syntax.

Book Overview

Chapter 2 Modeling and Simulation Background

This chapter gives a brief history of analog and digital simulation techniques using SPICE [2]. It also elaborates on digital and mixed-signal simulation technologies using HDLs, introduces both analog, digital and mixed-signal HDLs, and elaborates on the basic concept of abstraction for modeling. It also defines the modeling continuum, discusses model precision and accuracy, describes the different types of analog modeling, and concludes with a description of typical applications of analog behavioral modeling/macro-modeling.

Chapter 3 Methodology

This chapter begins with a description of common IC design methodologies and describes how modeling can play a role in these methodologies. The chapter concludes with a description of an analog behavioral modeling methodology.

Chapter 4 Primitive Building Blocks

This chapter begins with a mini MAST tutorial. It also describes the design and simulation of primitive-level analog behavioral modeling building blocks such as voltage arithmetic blocks, electrical sources, and basic devices.

Chapter 5 Functional Building Blocks

This chapter elaborates on the design and simulation of functional-level analog behavioral modeling building blocks such as a peak detector, a Schmitt Trigger, basic digital logic, a pulse-width modulator, an analog-to-digital converter, and a DC motor.

Chapter 6 System Examples

This chapter implements analog behavioral models/macro-models for analyzing system-level designs. Specifically, a distributed supply, an automotive ignition system, an audio test system, and a digital communications system will be modeled and analyzed.

Modeling and Simulation Environment

The Saber® computer-aided design tool was used for simulation, the DesignStar® schematic capture tool was used to create the schematics, and Pltool® was the graphics package used for displaying the results. The modeling examples were implemented in MAST, the analog hardware description language utilized by the Saber simulator. [3][4][5]

REFERENECES

[1] MAST is a registered trademark of Analogy Inc.

[2] *Simulation Program with Integrated Circuit Emphasis*, University of California Berkeley.

[3] Saber is a registered trademark of American Airlines, Inc., licensed to Analogy Inc.

[4] DesignStar is a registered trademark of Analogy Inc.

[5] Pltool is a registered trademark of Analogy Inc.

2

Modeling and Simulation Background

The reader will find this chapter useful for background on modeling and simulation. It sets the context with respect to analog behavioral modeling and simulation for the rest of the book. First, it discusses the different categories of simulation and the types of simulators within those categories. This chapter also elaborates on some specific HDLs as well as gives background on the development of standards that are evolving for analog and digital HDLs. It discusses the significance of abstraction for modeling in general, and defines the modeling continuum. This is followed by a discussion on model accuracy and precision. A description of the differences between some common modeling techniques for IC design is also given. The chapter concludes with a discussion on common applications of analog behavioral modeling.

2.1 Analog Simulation

Analog simulation entails detailed mathematical methods for simulation. Analog simulation usually involves the solving of non-linear differential equations via nodal analysis of the elements in a circuit. In electrical terms, two variables are defined at each circuit node: voltage (across variable) and current (through variable). The equations that define each node form a system of equations. These equations can be solved through Gaussian elimination or through some other algebraic technique. This type of simulation allows for a transient analysis of the dynamics within a circuit. Simulation Program with Integrated Circuit Emphasis (SPICE) simulation is one of the more common electrical simulators that utilizes analog simulation.

2.1.1 The SPICE Simulator

In the 1960's, at the University of California, Berkeley, the Integrated Circuit Group of the Electronics Research Laboratory developed SPICE. SPICE was later released to the public in 1970. Today, SPICE is one of the most widely used simulation programs for analog circuit design and is used to characterize gate-level models for simulation of complex digital systems. SPICE can be used to obtain DC operating points, transient responses, and frequency responses of electrical circuits.[1]

Over the years, SPICE has gone through many revisions. The most common version is SPICE2, in which the kernel algorithms were upgraded to support advanced integrated system methods, many of which relate to IC performance. The most recent version is SPICE3. In this version, the program was converted from FORTRAN to C for easier portability. In addition, several devices were added to the program library, such as a varactor, semiconductor resistor, and lossy RC transmission line models. However, the kernel algorithms have not changed and newly developed device models can still be simulated in SPICE2 using external device modeling techniques.[2]

Today, there are more than 35 SPICE derivative programs which have taken on custom acronyms. There are two flavors of SPICE: SPICE for the mainframe computer, and SPICE for the PC; HSPICE from Avant!, I-SPICE from NCSS Time Sharing, Z-SPICE from Z-Tech, and PSPICE®[3] from Orcad.[1] There are also many proprietary versions of SPICE, internal to semiconductor companies.

With the increased complexity of analog design, SPICE pushed CPU limits with respect to simulation time and disc space; with this limitation, macro-modeling/simulation with SPICE evolved. Macro-modeling is defined as

Analog Simulation

using pre-defined components to abstract the behavior of components at a higher level of abstraction.[2] Macro-modeling will be covered in more detail later in this chapter.

2.1.2 AHDL Simulator

Eventually, analog designs such as phase-locked loops, switching power supplies, and data converters become too large and complex for practical SPICE simulation; hence the advent of Analog Hardware Description Language (AHDL) simulators. The first AHDL based simulator for analog design was Saber, from Analogy, which was released in the fall of 1986. Saber initially had only analog modeling capability but was designed from the beginning for mixed-signal design and simulation; the mixed-signal capability was introduced in 1988. Other analog behavioral modeling tools include: Eldo®[4] from Anacad, SpectreHDL®[5] from Cadence, Verilog-A Explorer from Apteq Design Systems, and SMASH®[6] from Dolphin Design.

In an AHDL simulation, the simulator solves the system matrix at a specific point, based on the time-stepping algorithm. Usually, the time step is controlled by the fastest changing signal in the simulation. AHDLs that provide only continuous time definitions are limited because they can not model analog behavior in terms of analog states. For example, in a comparator the output voltage can be modeled by a linear transition from 0 volts to VCC - this is analog behavior. Modeling analog state behavior, on the other hand, is possible when the transition can be modeled by an instantaneous transition from 0 to VCC. This capability is beneficial when simulation time becomes a factor, especially when the models become very large or when simulations exhibit high-frequency switching.

See Figure 2.1 for a graphical representation of an AHDL simulator. The input is AHDL code containing both an analog netlist and models. The simulator then builds a sparse matrix. The equation solver uses iterative methods (e.g Newton-Raphson) to solve the nonlinear matrix. Convergence control verifies that KCL (i.e. energy conservation) is satisfied and that the solution is within tolerance before accepting the next time-step and moving on to the next time point.[7][8] Analogy's Saber and Anacad's Eldo were the first two analog AHDL simulators.

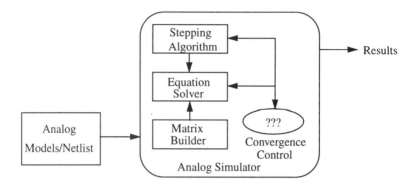

Figure 2.1 AHDL analog simulator.

2.2 Digital Simulation

Digital simulation can be performed through SPICE simulation, digital HDL simulation, or by using traditional programming techniques in C or Fortran. The traditional programming techniques are beyond the scope of this book.

2.2.1 The SPICE Simulator

The first digital ICs included simple logic functions such as gates, counters, registers, and adders that could easily be simulated in SPICE. These digital building blocks are technically analog circuits, and are still optimized today using SPICE.

Eventually, digital simulation via SPICE reached its limits. Digital circuits containing thousands of nodes could not be simulated in reasonable amounts of time. Another limitation was the amount of computer memory needed for simulation, exceeded what was practically feasible. These limitations created the need to further abstract circuit functionality by modeling digital behavior. Modeling digital behavior meant capturing functionality in terms of boolean expressions that resulted in digital output states (i.e. logic 1s and 0s). Hence the advent of digital HDL simulators which extended simulation capabilities into gate-level abstractions.

2.2.2 Digital HDL Simulators

Digital HDL simulators are usually event-driven. An event is considered to occur when there is a change of state in a node that may affect other components in a circuit. In an event-driven simulation, an event queue records all the states that will change at a particular time. See Figure 2.2 for a graphical representation of an event-driven simulator. [9] The most common digital simulators are Verilog and VHDL based simulators.

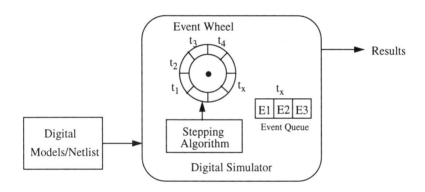

Figure 2.2 Digital event-driven simulator.

2.3 Mixed-Signal Simulation

Today, we are in an era of hyper-integration of IC systems where IC design is too complex for any one designer to manage. Boundaries between analog and digital systems are becoming non-existent. Integrating both analog and digital functionality is known as mixed-signal design. Mixed-signal design requires analog and digital simulation (i.e. continuous time analog simulation and digital even-driven simulation) for both design and verification. Mixed-signal modeling and simulation is an important step for the design integration of entire systems on to a single IC.

Mixed-Signal HDL-based simulators are the key for such simulation. Analog and digital behavioral modeling can be merged such that a true mixed-mode simulation can be used to verify functionality and connectivity of an IC system. Today, there are many products that are merging the analog and digital simulation. There are two categories of mixed-signal simulation technology: mixed-mode simulation and native mixed-signal simulation.[10]

2.3.3 Mixed-Mode HDL Simulator

Mixed-mode simulation is defined as the co-simulation of two separate simulators. The co-simulation consists of two simulation kernels connected by the time-stepping algorithms, a third kernel. See Figure 2.3 for a graphical representation of a mixed-mode HDL simulator.

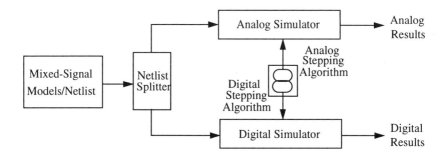

Figure 2.3 Mixed-mode HDL simulator.

One kernel contains the analog simulation, another kernel the digital simulation, and a third kernel performs the time-stepping algorithms; the analog and digital time-stepping mechanisms are separate. Time-steps are controlled by the hand-shaking between the two simulators (i.e. one simulator will not proceed until it has the needed data from the other simulator).

There are many types of algorithms that control the time-step synchronization between the analog and digital. The following table describes a few of the algorithms.[11]

Algorithm	Description
lock-step	smallest of the analog or digital step-size determines time-step
digital-control	digital simulator determines time-step
analog-control	analog simulator determines time-step
rollback	the analog and digital simulators each determine their own time-step; synchronization occurs when there is communication between the two domains - the analog simulator "rolls back" to the time where the interaction occurred

A disadvantage of this type of technology is the lack of error-checking between the analog and digital simulation to assure the data makes sense before continuing either of the simulations; each simulator will proceed even if the data is erroneous. Another disadvantage is that two separate simulators need to be purchased in order to perform mixed-mode simulation. Furthermore, the three separate kernels require more computer memory and simulation time as opposed to a single kernel.

Advantages of this technology are existing models for digital designs can be immediately used in the co-simulation and then directly synthesized after the design has been verified. An example of mixed-mode simulation is the co-simulation of Saber and Verilog.

2.3.4 Native Mixed-Signal HDL Simulator

Native mixed-signal simulator technology allows both the analog and digital simulation to occur under a single kernel. Another characteristic is that the time-stepping mechanisms are connected and work together via a mixed-signal algorithm. This type of technology enables error-checking between the analog and digital simulation. Moreover, a uniform debugger can be implemented allowing the user to debug the analog and digital portions of a model simultaneously.[8]

See Figure 2.4 for a graphical representation of a native mixed-signal HDL simulator. In a native mixed-signal simulation, the netlist is split between the analog and digital as is done in the mixed-mode simulator. The interaction between the simulators occurs via a mixed-signal time-stepping algorithm which verifies consistency between the analog and digital simulator outputs before continuing on to the next time-step.[12]

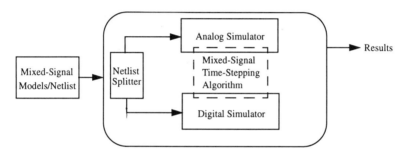

Figure 2.4 Native mixed-signal HDL simulator.

An example of a native mixed-signal simulator is Saber from Analogy. This simulator uses a mixed-signal AHDL (MAST) where both analog and digital (event-driven) functionality can be defined in the same model.

Now that we understand how different types of simulators function, we can appreciate the higher level languages that these simulators utilize to describe mixed-signal behavior.

2.4 HDL Basics

A hardware description language (HDL) is a higher level language that can describe component behavior as well establish connectivity (i.e. netlist). HDLs enable many levels of abstraction. These multi-levels of abstraction can then be simulated simultaneously. HDL models are user defined and linked to

HDL Basics

a simulator. There are HDLs designed specifically for both analog and digital IC design. This section elaborates on HDLs available for both analog and digital design. It also discusses the differences between analog HDLs and mixed-signal HDLs.

2.4.1 Digital HDLs

The evolution of HDLs in the digital domain was the result of the increasing complexity of digital ICs. Every digital designer has a fundamental understanding of a digital gate at the transistor level. This understanding allows the designer to abstract the transistor level detail by a digital boolean function. Through gate level abstractions (i.e. AND gates, flip-flops, latches, etc.), larger designs can be analyzed without incorporating extraneous detail. Continuing through higher levels of abstractions, the gate-level can be abstracted even further: state machines, registers, counters, combinational logic, etc., can be used to design even more complex systems without incorporating detail at the gate level. For example, designing CPUs, encoders, DSP controllers, I/O interfaces, would be impractical to design initially at the transistor level or at the gate-level. System-level design issues need to be worked out before commencing with detailed design.

HDLs were developed to support top-down design methodologies. Through digital HDLs, design can become an automated process by defining systems at the top-level via behavioral descriptions and then by synthesizing those descriptions down to the gate level. Hence, the name digital synthesis.

There are two digital HDL standards that have become widely accepted throughout the world: Verilog and VHDL.

VHDL

In the 1970's the Department of Defense (DoD) began to use a programming language specifically designed to describe electronic hardware, also known as HDL. The HDL took a different approach from SPICE by abstracting digital functions using higher level programming language constructs to focus on the behavioral aspects of the design, rather than on the gate or transistor behavior.

In the 1980's, the DoD's HDL evolved and expanded to include very high-speed IC designs (VHSIC) and changed the language name to VHDL (VHSIC Hardware Description Language). By 1985, the language was released for public review. The primary objective of the language was to provide a vehicle for improved documentation of electronic systems delivered to the government.

In 1986, all rights to the VHDL language were transferred to the IEEE as part of an industry-wide standard for a hardware description language. This standardization of VHDL was driven by the EDA industry and users of the design tools. VHDL became an IEEE standard in 1987 and was officially designated IEEE 1076 in 1988 and has since been revised by the IEEE in 1993.[13]

VHDL was considered one of the most important HDLs by the military because the standardization eliminated design risk that existed with proprietary languages. VHDL is considered to be ADA-like in syntax. Companies bidding for military contracts had no choice but to adopt the VHDL standard in order to be competitive. Internal government standards required that every physical implementation level of an electronic system be documented and contain both a structural netlist and an executable behavioral description. These requirements included a high-level behavioral model for the entire system, for each board and every chip in the system. Unfortunately, VHDL encompassed only digital electronic systems and left the analog counterparts to be documented using SPICE. These standards were adopted by the military to capture the essence of the design without exposing low-level design detail. Furthermore, this documentation process would also support top-down design methodologies.[13] See Example 2.1 for a VHDL implementation of an AND gate.

```
entity And is
 generic (td: time: = 1.0 ns);
 port(X,Y: in bit;
    Z:   out bit);
end AND;
architecture A1 of AND is
begin
    z <= X and Y after td;
end A1;
```

Example 2.1 VHDL AND gate.

Verilog

Verilog is another HDL that provides behavioral modeling capability of digital constructs at many different levels of abstraction. Verilog was developed in 1983 and 1984.[14] Verilog has become widely used in the United States while VHDL has become more popular in Europe.

A organization known as Open Verilog International (OVI) has since been formed to promote the use of Verilog. The organization is responsible for soliciting input from the industry to support the Verilog language. Verilog has successfully achieved standardization through balloting IEEE in December of 1995.

Verilog is another example of a HDL created for digital IC design. Verilog was released as a proprietary language in 1983. It has since become widely used. See Example 2.2 for a Verilog implementation of an AND gate. The Verilog implementation is very similar to the VHDL implementation in that the model name and ports are defined first, the and boolean function second.

```
module and(Z,X,Y);
input X,Y;
output Z;
 assign Z = X & B;
endmodule
```

Example 2.2 Verilog AND gate.

2.4.2 Analog HDLs (AHDLs)

As with the digital entities, there exist well-understood analog constructs (i.e. resistor, capacitor, comparator, operational amplifier, peak detector, etc.). The device-level equivalents of these constructs can be abstracted to design even higher-level blocks. AHDLs make this abstraction possible in the analog domain, as was done with the digital HDLs in the digital domain. It should be noted that AHDLs can still be utilized to model analog behavior at the device-level; unlike SPICE models, however, AHDL models can easily be changed since they are not pre-defined.

There are two categories of AHDLs: those that can define only analog constructs and those that can define both analog and digital constructs together. An AHDL that can define only analog constructs describes behavior in terms of mathematical equations and establishes connectivity as in a SPICE netlist.

In an analog-only HDL, the across and through variables (e.g. current in a current source and voltage in a voltage source) are determined only at the inputs and outputs of an analog component. The behavior of an analog component is described in terms of an equation, thereby reducing the number of electrical nodes needed to define a system of analog components. In other words, the system of equations defined in the simulator is minimized because the analog behavior of the various blocks within the system is defined by equations, as opposed to the many nodes of the elements (i.e. resistors, capacitors, transistors, etc.) that those blocks contain.

Digital behavior, on the other hand, can still be described in terms of equations that define current and voltage nodes as inputs/outputs, as opposed to describing behavior in terms of digital events (i.e. 1s and 0s). There are many languages that utilize analog-only HDLs:

ABDC

ABCD is an AHDL that is has been designed for analog systems. It is derived from a combination of SPICE and C constructs. This language has been adopted by Dolphin Integration, developer of the SMASH mixed-signal simulator.[15] Example 2.3 shows ABCD model code for a resistor.

```
.SUBCKT RESISTOR P M
+ PARAMS: res = 1e3
 VPM P M <behavioral>
[behavior]
 VPM~val = res * VPM~current;
.END RESISTOR
```

Example 2.3 ABCD resistor AHDL model code. [16]

ABCD is unique in that it is SPICE-like because it uses the familiar *SUBCKT* SPICE construct.[16] For further information on ABCD and SMASH please see following internet home page:

http://www.dolphin.fr/faq.html#abcd.

SpectreHDL

SpectreHDL is an AHDL that is designed for analog systems. SpectreHDL is a proprietary language developed at Cadence Design Systems. This language was developed to work with Cadence's Spectre suite of simulators. SpectreHDL was derived from Verilog and C constructs. SpectreHDL can be used to define models in the electrical, mechanical, thermal and fluid system domains. Example 2.4 shows SpectreHDL model code for a resistor.

```
module resistor(p,m) (res)
node [V,I p,m;
parameter real res = 1.0 from (0:inf);
{
 analog{
    I(p,m) <- V(p,m)/res;
}
}
```

Example 2.4 SpectreHDL resistor model code. [17]

For further information on the SpectreHDL modeling language see the following internet home page:
http://www.cadence.com.

Verilog-A

Verilog-A is an AHDL that is designed for analog systems. Verilog-A is an extension to the IEEE 1364 Verilog HDL specification. The Open Verilog International (OVI) committee developed the specification for Verilog-A in 1996. The specification for Verilog-A will eventually become a subset of Verilog-AMS and reviewed by IEEE with the intention that it will become a standardized extension of the current Verilog specification. Example 2.5 shows Verilog-A model code for a resistor.

```
module resistor (p,m);
inout p,m;
electrical p,m;
```

17

```
parameter real RES = 1.0;
 analog
    V(p,m) <+ RES*I(p,m);
endmodule
```

Example 2.5 Verilog-A resistor model code. [18]

The syntax of this example is likely to change since IEEE has not yet approved the Verilog-AMS language specification. Verilog-A is currently being adopted by various design automation vendors.

2.4.3 Mixed-Signal HDLs

A mixed-signal HDL can define both analog and digital constructs in terms of mathematical equations, boolean expressions, and logical assignments and establish connectivity as in a SPICE netlist. In a mixed-signal model, both analog and digital (event-driven signals) are defined in the same model; analog behavior can be described in terms of equations and the digital behavior in terms of digital events (i.e. 1s and 0s). This implies that a mixed-signal language must provide constructs that allow communication between the analog and digital domains. Furthermore, because digital behavior can be described by events, the analog portion of the simulator does not have to solve a system of equations that describe digital behavior; thus the amount computer memory and time needed for simulation of a large mixed-signal system is reduced.

The best mixed-signal AHDLs provide the robustness needed for users to have more flexibility in defining models. AHDLs such as MAST, HDL-A, are a few of the *de-facto* standard mixed-signal languages currently used in industry. However, these languages are proprietary languages which make it difficult to use a single model on more than one simulator. For this reason, the development of standard languages will eventually make it possible to exchange models without having to worry about simulator compatibility. However, such languages are not yet completed. They are still undergoing the task of becoming standards. The following are examples of some AHDLs available today:

VHDL-AMS

VHDL-AMS is a mixed-signal AHDL that has been designed for analog and digital systems. The VHDL-AMS language specification was developed within IEEE in order to become a standardized language. VHDL-AMS is not an extension of the VHDL 1076-1993 specification; 1076-1993 will become a subset of the VHDL-AMS specification. The VHDL-AMS language specification is being developed by the IEEE 1076.1 working group. Further information can be found regarding the IEEE 1076.1 working group on the following internet home page:

http://www.vhdl.org/vi/analog/.

Example 2.6 shows VHDL-AMS model code for a resistor.

```
ENTITY resistor IS
 GENERIC(res:REAL:= 1);
 PORT(p,m: electrical);
END resistor;

ARCHITECTURE basic_res OF resistor IS
BEGIN
 RELATION BEGIN
 [p,m].I <+ [p,m].V/res;
 END RELATION;
END ARCHITECTURE basic_res;
```

Example 2.6 VHDL-AMS resistor model code. [19]

In VHDL-AMS mixed-signal models can be defined. Since the syntax is not mature, no further examples in this language will be shown.

Verilog-AMS

Verilog-AMS is a planned mixed-signal AHDL standard that will evolve from the Verilog and Verilog-A language specifications. Verilog-AMS will consist of the Verilog and Verilog-A languages linked together by a mixed-signal interface. [20]

HDL-A

HDL-A®[21] is a mixed-signal AHDL that has also been designed for analog and digital systems. HDL-A is a proprietary language developed at Anacad of Mentor Graphics Corporation. This language was initially developed to work with Anacad's Eldo simulator. This language has now been expanded to work with the AccuSimII simulator of Mentor Graphics. HDL-A was derived from the VHDL-AMS language specification. However, it not fully compatible with VHDL-AMS. HDL-A also supports more than electrical system domains; it can support mechanical, thermal and fluid systems. Example 2.7 shows HDL-A model code for a resistor.

```
ENTITY resistor IS
 GENERIC(res: ANALOG);
 PIN (p,m: ELECTRICAL);
END ENTITY resistor;

ARCHITECTURE basic_res OF resistor IS
BEGIN
 RELATION
    PROCEDURE FOR DC, TRANSIENT, AC =>
       [p,m].i%= [p,m].v/res;
 END RELATION;
END ARCHITECTURE basic_res;
```

Example 2.7 HDL-A resistor model code. [22]

Because HDL-A is also a mixed-signal HDL, both analog and digital constructs can be defined in the same model. The characteristics of a simple comparator can be mixed-signal in nature and would be a good example to demonstrate a mixed-signal model. See Example 2.8 for the HDL-A model code for a comparator.

```
ENTITY cmp IS
   GENERIC (vth : analog; td : time);
   PORT (cmpout: inout BIT);
   PIN  (in1,in2: electrical);
END cmp;
ARCHITECTURE level1 OF cmp IS
```

HDL Basics

```
BEGIN
  PROCESS
    BEGIN
      IF ([in1,in2].v >= 0.0) THEN
        cmpout <= '1' ;
      ELSE
        cmpout <= '0' ;
      END IF;
    LOOP
      WAIT ON rising([in1,in2].v, vth),
              falling([in1,in2].v, vth) ;
      IF rising([in1,in2].v, vth) THEN
        cmpout <= '1' after td ;
      ELSE
        cmpout <= '0' after td ;
      END IF ;
    END LOOP ;
  END PROCESS ;
  RELATION
    PROCEDURAL FOR init =>
      vth := 0.0
      td := 10 ns ;
  END RELATION;
END ARCHITECTURE level1;
```

Example 2.8 HDL-A model code for a comparator.

This model has analog inputs *in1* and *in2* and digital output *cmpout*. The model determines whether or not the differential input voltage is positive or negative and sets the digital output to the respective value.

For further information on the HDL-A modeling language see the following internet home page:

http://www.mentor.com/hdla/datasheet.html.

MAST

MAST is a mixed-signal AHDL that was also designed for analog and

21

digital systems. MAST is a proprietary language developed at Analogy Inc. This language was developed to work with Analogy's Saber simulator. MAST was the first mixed-signal AHDL and therefore was not based on any standardized language - it is considered to be a *de facto* standard, especially in the automotive electronics industry.[10] MAST also supports more than just electrical system domains; it can support mechanical, thermal and fluid systems. Example 2.9 shows the MAST model code for a resistor.

```
template resistor p m = res
 electrical p,m
number res
{
equations{
 i(p->m) += (v(p)-v(m))/res
   }
}
```

Example 2.9 MAST resistor model code. [23]

MAST also has the capability to model the mixed-signal behavior of a comparator. See Example 2.10 for the MAST model code for a simple comparator.

```
template cmp in1 in2 out = td
electrical in1,in2
state logic_4 out
number td=0
{
val v vdiff
state nu before,after
when(dc_init) {
   schedule_event(time,out,14_0)
            }
when(threshold(vdiff,0,before,after)) {
    if((before == -1) & (after == 1)) {
        schedule_event(time+td,out,14_1)
                                      }
    else schedule_event(time+td,out,14_0)
```

Abstraction

```
         values{
            vdiff  = v(in1) - v(in2)
                }
         }
                                                                   }
```

Example 2.10 MAST model code for a comparator.

This model has analog inputs *in1* and *in2* and digital output *out*. This model compares the differential input voltage with 0 and sets the digital output to the respective value.

For further information on the MAST modeling language see the following internet home page:

`http://www.analogy.com/products/prod.htm#Saber.`

This book will use the MAST AHDL to implement the model examples since it is considered one of the more mature languages and has been used in the industry for the longest period of time.[10]

Having explored a few of the high-level languages that make it possible to model analog and digital behavior, we can now embark on an understanding of how the concept of abstraction can help us to analyze IC system behavior in an efficient manner with respect to simulation.

2.5 Abstraction

Abstraction is one the most powerful engineering design concepts. There are abstractions of data, software, hardware, ideas, concepts, and theories that enable the mind to explore vast engineering problems. Abstraction is an engineering tool by which complex systems can be broken down into subsystems whose behavior can be precisely defined and understood. Examples of mathematical abstractions are the trigonometric functions on a calculator. Today, students are not taught the underlying details of the actual calculation of the sine of 30 degree; a student takes for granted that 0.5 is the answer. The sine of 30 degrees can then be used to calculate the answers to more complex problems.

In the same manner, an IC designer can abstract the behavior of subsystems to build larger systems. In other words, concisely defined subsystems can be used to define higher level systems. Unlike the student, a designer should not take abstraction for granted; experience and an understanding of the underlying concepts are the key to concise abstraction.

Specifically, in IC design, abstraction of device-level behavior is often a concern because it is difficult to understand what information has been neglected for higher-level abstractions. It is important to focus on the details that are necessary to obtain the results that can reveal the critical problems. For example, in modeling the functional behavior of a bandgap reference, it is sufficient to only consider the range of supply voltage that will guarantee that the underlying transistors are in the linear region (i.e. not saturated). In this case, the bandgap model needs only to verify that the supply voltage is at a reasonable level during all time; modeling device-level equations would give too much detail for what is expected from the functional model. Lower levels of simulation (e.g. SPICE simulation), however, can be used to characterize the range of supply input that would maintain nominal operation. The characterization data can then be used to specify the value of the parameter in a functional-level model that defines this range of operation.

As a second example, let us consider an operational amplifier. In abstracting an operational amplifier it is important to consider the device-level characteristics which determine higher- level behavior. But this does not imply that device-level detail must be incorporated into a higher-level model. A device-level model uses device parameters (e.g. IS, BETA, CJS, VAF) in bipolar devices. A higher-level model characterizes an operational amplifier in terms of: first pole, second pole, offset voltage, slew rate etc. These characteristics can be determined through actual measurement in the laboratory or through SPICE simulation. Hence, higher-level equations (not SPICE equations) that define behavioral models in terms of the poles, offset voltage, and slew rate, can be used to abstract the operational amplifier, while maintaining the important characteristics. Figure 2.5 illustrates the connection between a functional-level operational amplifier description to the underlying device characteristics.

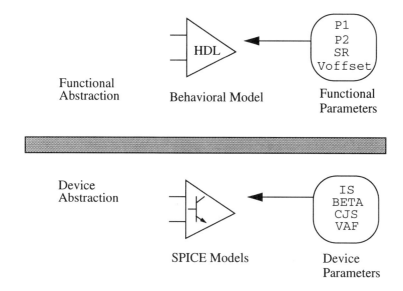

Figure 2.5 Functional and device-level abstractions of an operational amplifier.

Having some exposure to the concept of abstraction, we can now proceed to establish various levels of abstraction that will enable us to concisely define the behavior of large IC systems.

2.6 Modeling Continuum

The modeling continuum is a defined hierarchy of abstractions that tie higher-level models to lower-level models. See Figure 2.6 for a graphical representation of the modeling continuum. The continuum can be thought of as three-dimensional space with the following axes: model detail, simulation time, and IC complexity.

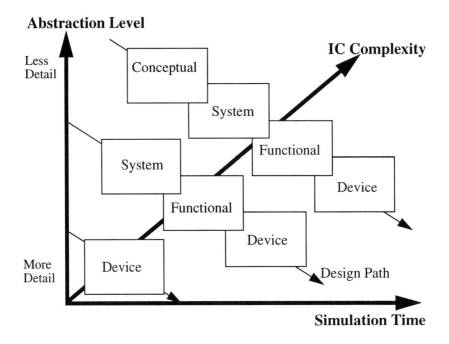

Figure 2.6 Modeling continuum.

There are four basic abstraction levels:

Device-Level

Device-level models are implemented at the transistor level (e.g. MOS, Bipolar, BiCMOS).

Functional-Level

Functional-level models capture critical functions of the blocks within in a system which are crucial to the operation of the system (e.g. gain blocks, amplifiers, peak detectors, phase-lock loops, pulse-width modulators, etc.). There can be many levels of this type of abstraction.

System-Level

System-level models contain the behavior of specific topologies or architectures of an IC system. Examples of different power supply topologies include: linear regulators, buck switching regulator, boost switching regulator, flyback switching regulator, etc.

Conceptual-Level

Conceptual-level modeling captures behavior at the highest abstraction level. At the conceptual-level, it is not important to include implementation detail. However, it is important to include algorithms that describe the concept. For example, a DSP algorithm is a conceptual design; how the design is implemented is a different story. Conceptual-level models are usually implemented in high-level description languages (HLDs) (e.g. MATLAB®[24], Mathcad, SPW®[25], programming languages, etc.).

These defined abstraction levels are arbitrary; there is no exact number of abstraction levels. The important thing to remember is that through abstraction we can build higher-level constructs that enable the definition of larger systems. Less complex IC designs require less abstraction levels to complete and verify designs. As IC designs become more complex, more abstraction levels are needed to completely understand and verify the design of an IC at the system-level. Therefore, depending on the complexity, there will exist different design paths - each path requiring more or less detail. Overlap between abstraction levels helps bridge the gap between higher and lower levels of circuit design. This can be done by extracting the important second-order information from the more detailed models and incorporating this information in the higher-level models. Hence, there is a seamless design path from concept to device implementation.

There are many different categories of simulation tools that span the modeling continuum. Some tools can be used in more than one level of abstraction. Examples of tools that can be used in the higher levels of abstraction are the fixed time-step tools (e.g. Matlab, SPW, or even C programming). Lower level abstraction tools are usually those tools that model components at the device-level (e.g SPICE). Analog behavioral modeling tools can be used in the system, functional, or device abstraction levels.

There is no exact method for using the various simulation tools for design.

The key concept to remember is that more abstraction (less model detail) implies less simulation time while less abstraction (more model detail) implies more simulation time.

This book will focus only on those levels of abstraction that are within the scope of analog behavioral modeling.

2.7 Model Precision and Accuracy

Model accuracy is related to simulator accuracy. An accurate simulator should produce accurate results with respect to the models. For example, if the model specified for a resistor is a straight line, then simulated behavior should also be a straight line. In Figure 2.7, two different simulation results are shown for the same resistor model. The current through the resistor is described by the y-axis (I) and the external voltage applied across the resistor is the x-axis (V).

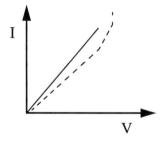

Figure 2.7 Model accuracy comparison of simulated I-V curves for a resistor.

The solid line is considered to be an accurate simulation result of what was defined in the model (a straight line). The dotted line is not an accurate simulation result of the defined model.

There are many reasons why a simulation result can become inaccurate. One reason is that sometimes a simulator can take too large time-steps for transitions defined in the model; this can be due to loosely defined accuracy settings in the simulator. Other reasons can be due to: mathematical overflow or underflow, across or through variable discontinuities defined in a model. In some cases, transitions of an across or through variable from one value to another should be performed in a continuous manner, as opposed to an instantaneous step function. Discontinuities can cause problems with analog simulator convergence and give inaccurate results.

There are, however, trade-offs between model precision and the amount of detail incorporated in a model. In Figure 2.8, two simulated I-V curves of different diode models are shown. The current through the diode is described by the y-axis (I) and the external voltage applied across the diode is the x-axis (V_A).

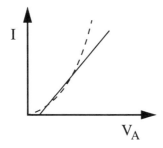

Figure 2.8 Model precision comparison of simulated I-V curves for a diode.

In this example, the dotted line simulation result is a more precise abstraction with respect to actual diode behavior, than the solid line result. More detail or information is described by the dotted line model. Both results, however, are accurate for their respective model definitions.

A more precise model is a model that contains more effects (less abstraction) so that the model more closely emulates the actual component behavior. Models that contain more abstraction do not necessarily compromise accuracy. As long as the simulation results reflect the behavior defined in the model, then the model is considered to be accurate with respect to the model definition. In general a model is only as good as its own definition. If a model is not defined correctly then good results can not be expected. A model will not tell that your design is guaranteed to work. However, it will tell you that your design will not work. Modeling is not a crutch to make a bad designer a good designer. Although, it will help a good designer become more efficient.

Now that we have explored the basic concepts of modeling, we can further explore the specifics of IC modeling.

2.8 IC Modeling

IC component behavior can be abstracted through various types of modeling. The EDA industry has made it possible to simulate these models by using complex algorithms that make millions of calculations in small amounts of time. What would take years to calculate on older computers can now be done in minutes using faster computers and efficient simulation tools. Simulation technology has made it possible to model and simulate DC, AC, and transient conditions of large IC systems. Moreover, noise and statistical analyses can also be performed. Designs can be optimized through sensitivity analyses. Despite the advances in modeling/simulation technology, modeling and simulation can not replace design. However, these tools can be used to aid the process of design.

To eliminate confusion, a definition for various types of modeling in the context of this book are given. An example of an adder function will be used to demonstrate each type of modeling.

2.8.1 SPICE Device Modeling

SPICE modeling uses device models that are characterized at the device-level (e.g. Beta, Vsat, and Isat for a bipolar device). The device models implemented in SPICE are usually very complex. For example, the MOS model has 3 different levels of model detail that can be specified. This level of complexity usually requires a lot of simulation time and computer memory for large circuits. Every device has unique behavior and that behavior is specified by equations. The equations for the devices are usually embedded in the simulator and are transparent to the user. See Equation 2.1 for an example of an equation for the DC operation of the collector current in bipolar transistor. This equation is an Ebers-Moll equation that is commonly used in SPICE programs. [26]

IC Modeling

$$IC = I_{ES}\alpha_F\left[\exp\left(\frac{VBE}{Vt}\right) - 1\right] + I_{ES} \qquad \text{(Eq 2.1)}$$

Various devices can be defined in a netlist to form a circuit and then be simulated. See Example 2.11 for a sample netlist of a Darlington pair amplifier. This particular netlist specifies: connectivity, parameters for a bipolar device, and the plotting specification.

```
VCC 2 0 DC 10V
VIN 1 0 CD 5V
* Device Instantiation
Q1   2   1 3 QMODEL
Q2   2 3 4 QMODEL
RB 2 1 50K
RE 4 0 5K
* BJT Model Parameters
.MODEL QMODEL NPN (BF=100, BR=1, RB=10, RC=1,
 RE=0,VJE=0.9,VA=100)
* Transient Analysis from 0 to 2ms by.1us
.TRAN 0.1US 2.0MS
* Plotting Specification
.PLOT TRAN V(1) V(2) V(3)
.END
```

Example 2.11 SPICE netlist for a Darlington pair amplifier.

Schematic capture programs can be used as graphical interfaces through which netlists are automatically generated. See Figure 2.9 for an example of an adder function modeled at the device-level.

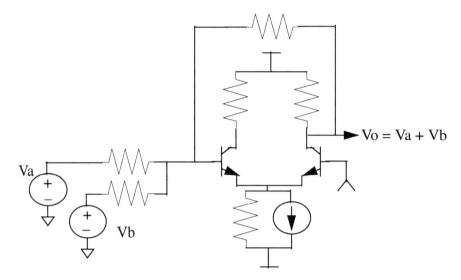

Figure 2.9 Device model implementation of an adder function.

From this schematic, a SPICE netlist similar to the netlist shown for the Darlington pair amplifier can be created. In cases where the circuits are large, the netlist can become very long.

2.8.2 SPICE Macromodeling

As with digital designs, the size of analog circuit designs has also become too large to practically simulate in SPICE. Macromodeling, a higher means of abstraction that took advantage of the SPICE simulator, was developed. Macromodels were introduced in 1974.

Macromodeling uses pre-defined components (controlled sources, resistors, capacitors, etc.) to define behavior. In macromodels, functionality is defined by a mathematical relationship that is performed through controlled voltage and current sources. Through this method, voltage/current addition and subtraction can be performed using parallel current sources and series voltage sources. Division and multiplication are performed using resistors; differentiation and integration are performed using capacitors and inductors.

IC Modeling

Macromodels are easy to use when describing simple behavior. [2] However, when the behavior becomes complex, defining functionality in terms of basic circuit elements can be difficult and sometimes impossible. See Figure 2.10 for an example of a macromodel implementation of an adder function.

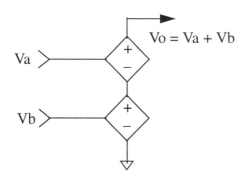

Figure 2.10 Macromodel implementation of an adder function.

2.8.3 Analog Behavioral Modeling

Analog behavioral modeling (ABM) utilizes an analog hardware description language through which behavior is described. ABM allows the user to define procedures that describe behavior as a functions of circuit variables (voltage, current, simulation time, etc.). This type of modeling is sometimes confusing because SPICE equations are behavioral models. However, in SPICE the equations are embedded in the simulator, whereas a behavioral simulator will contain the equations for the devices in the AHDL. AHDL simulators can be just as accurate as SPICE simulators. However, since SPICE simulators are specialized for device-level modeling; they can be more efficient in device-level simulation.[10]

SPICE simulators can also be extended to model analog behavior. They can be extended by the following methods: use of polynomial controlled sources, modification of the simulator code to add new models, or by building macromodels. The polynomial method is good if the electrical behavior can be curve-fitted. The disadvantage of this approach is that a new curve has to be derived every time the underlying design is changed. Adding a new model to a SPICE simulator is very difficult because the underlying simulator code is complex. Also, adding new code makes the simulator non-standard. Macro-models can become complex when mathematical behavior is synthesized by electrical components (e.g. using a diode to perform a log function). [27]

Because ABM models are separate from the simulator, any behavior that can be described analytically can be modeled. Thermal and power calculations can also be modeled. Essentially, any across or through variable can be described through an ABM. Moreover, currents and voltages (across and through variables) have to be defined only at the inputs and outputs of a model. This can eliminate extraneous rows and columns in the system matrix. In the adder example, the output can be described by simply adding both input voltages. See Figure 2.11 for a graphical representation an analog behavioral model for the adder function. Note all internal nodes have been eliminated. Thus, the computation time in a simulation can be reduced as well as the computer memory needed to store the simulation data.[28]

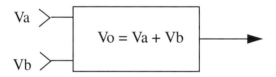

Figure 2.11 Analog behavioral model representation of an adder function.

2.8.4 Analog Behavioral Macromodeling

Analog behavioral macromodeling is similar to macro-modeling in that models are assembled by using existing modules. However, in behavioral macro-modeling the user is not limited to built-in modules - if it does not exist, it can be created. See Figure 2.12 for a graphical representation of an analog behavioral macro-model for a model that performs the addition of four voltages. This model can be constructed from the previously defined adder function.

IC Modeling

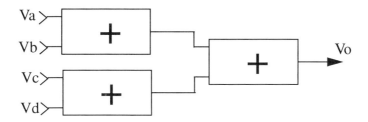

Figure 2.12 Analog behavioral macromodel representation of a four-input adder function.

2.8.5 Algorithmic Modeling

Algorithmic modeling is the most fundamental type of modeling in that this type of modeling does not involve simulations that maintain time-step control (e.g. fixed time step simulation). The most basic form of algorithmic modeling is done through equations that describe behavior in languages such as BASIC, C, or Fortran. In this type of modeling there is a fixed time step that can be incremented by looping through an equation. Modeling of this type has been practiced for many years. Algorithmic modeling is the fastest type of modeling with respect to simulation time. See Example 2.12 for an algorithmic model implementation for the adder function in C++. Other algorithmic modeling tools include: MATLAB, Mathcad, and SPW. These more advanced tools have evolved HLD languages. HLD languages are high-level languages that contain advanced constructs that make it easy to define conceptual behavior.

```
// C++ Algorithmic Program for an Adder
//Function
#include <iostream.h>
#include <stdlib.h>
void main()
{
  float vin1, vin2, out, x, y;
```

35

```
int t, t = 0;
for (t = 0, inc <= 1000, t = t+1)
{
x = 2 * 3.14 * t;
y = x + 3.14;
vin1 = sin(x);
vin2 = sin(y);
out = vin1 + vin2;
cout << "\n Vout =" << out << "time ="
     << t << "seconds";
}
}
```

Example 2.12 C++ adder function model code.

Now that we have gained a broad knowledge of various types of modeling for IC design, we can focus on ABM. Before getting into the details of the modeling languages it is important to understand how ABM can help us in the overall picture of product development.

2.9 Applications of Analog Behavioral Modeling

There are many applications for ABM. ABM can be used in every aspect of new-product development as illustrated below.

2.9.1 Dynamic Specification

Analog behavioral modeling can be used as a vehicle for defining new applications for existing ICs and developing new IC's through dynamic specification. Traditionally, all specifications are written on paper for systems engineers and IC designers to interpret. Specifying an IC through a model can be beneficial to both the customer and the IC designer. The development of IC systems can be verified for compatibility with the customer's application.

For example, in Figure 2.13 a system is shown where an IC specification can be derived without the actual hardware.

Applications of Analog Behavioral Modeling

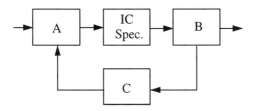

Figure 2.13 Dynamic specification.

Blocks A, B, and C may or may not yet exist. However, a model for these blocks can be designed so that the specification for another block can be derived from simulation. Initially, the model can be designed to be very simple. As more is known about the customer's application, more information or detail can be modeled. Eventually, an entire specification for an IC can be derived by just developing a model.

The advantage of this technique is that changes can be easily implemented and verified at a high-level in the entire system design.

2.9.2 System Design

Using ABM for system design can also be very beneficial. The main components of an IC system can be specified and understood as to how each individual block can affect the overall system. ABM can be used to develop the system architecture or for optimizing system performance. Stability issues can be explored in both the frequency and transient domains. Moreover, the analog and digital components can be simulated together to understand how mixed-signal designs can affect system performance.

Another system design application is using existing standard cells to implement a mixed-signal system. This concept is also known as reuse. Models of existing components can be used to implement and verify the system.

2.9.3 IC Design

For large IC designs, SPICE simulation will eventually require too much CPU time and memory. When SPICE simulation no longer becomes practical, ABM can be used to complement SPICE simulation. The more crucial blocks can be simulated at the device-level and verified in the overall system. This is made possible because ABM allows the user to mix analog behavioral models of higher level blocks with device-level models in a single simulation. Essentially, the nature of AHDLs allows the user to specify any level of abstraction in a single simulation (i.e. multi-level simulation).

Another application of ABM in IC design is verifying basic functionality and connectivity at the system-level before sending the design to the mask shop. ABM and simulation will not discover every design error. However, ABM is intended to reveal the gross errors. Ultimately, ABM will help to seamlessly integrate the system to the circuit design. The goal is to reduce the number of design iterations. Reducing the number of design iterations will save in the manufacturing costs, decrease the time for design and debugging, thereby, decreasing the time-to-market.

More examples of IC systems are shown in Chapter 6.

2.9.4 Virtual Test

ABM can be used to develop the test program for an IC. Through full-chip simulation, critical nodes can be identified for debugging and test pads can be put in the appropriate places. Design circuitry can also be designed and verified in the top-level simulation. One of the more powerful aspects of AHDLs is that measurement models that can calculate power, rise/fall times, slew rates, and duty cycle, without introducing error into the system, as would be the case in the laboratory. Finally, a model of the test fixture and code can be developed that will enable the test engineers to anticipate test problems before the test program is developed and the actual IC has been manufactured. See Figure 2.14 for a virtual test setup.

Applications of Analog Behavioral Modeling

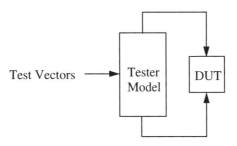

Figure 2.14 Virtual test setup.

The test setup model includes the test vectors, tester model, and the Device Under Test (DUT). The tester model would exercise the DUT as would the actual tester. This can save the amount of time spent on expensive test equipment for debugging test programs. Today there are ongoing research efforts in the area of virtual test. Teradyne, Credence, Hewlett Packard, and LTX, are a few of the Automatic Test Equipment (ATE) manufacturers that have virtual test products currently under development. For more information see the following internet homepage:

`http://www-new.cadence.com/dantes/VT_partners/VT_Partners.html`.

And for a white paper regarding simulated test to reduce time-to-market please see:

`http://www.virtualtest/dloads/isdapr97.html`.

2.9.5 Other Applications

Analog behavioral models can be used to develop electronic data sheets. The customer can enter the design requirements and the external values that setup the IC will be calculated automatically. For example, a designer can specify a desired input and output voltage of a DC-to-DC converter and the electronic data sheet would calculate the optimal compensation values for the external resistors and capacitors. Another application is that customers may choose to use characterized IC model libraries to develop their own applications. Most AHDLs offer encryption schemes that will keep the proprietary underlying design information confidential. This is especially useful when supplying libraries of models to customers.

As the standard languages evolve, and computers become more powerful there will be many other opportunities for ABM to assist in the product development process. Analog synthesis may eventually come to fruition using advanced ABM techniques.

This book focuses only on those applications related to IC system design.

REFERENCES

[1] Muhammad H. Rashid, *SPICE For Circuits And Electronics Using PSpice*, Prentice Hall, Englewood Cliffs, New Jersey, 1990.

[2] Connelly, Alvin, J, and Choi, Pyung, *MacroModeling with SPICE*, Prentice-Hall Inc., New Jersey, 1992.

[3] PSPICE is a registered trademark of MicroSim Corporation.

[4] Eldo is a registered trademark of ANACAD Computer Systems GmbH.

[5] SpectreHDL is a registered trademark of Cadence Design Systems, Inc.

[6] SMASH is a registered trademark of Dolphin Integration.

[7] Laurence W. Nagel, "SPICE2 A Computer Program to Simulate Semiconductor Circuits, College of Engineering, Memorandum No. UCB/ERL M520", University of California at Berkeley, Berkeley CA, May 1995.

[8] H. Alan Mantooth and Mike Fiegenbaum, *Modeling With An Analog Hardware Description Language*, Kluwer Academic Publishers, Boston, 1995.

[9] Cadence, *Verilog-XL Training Manual Version 2.0*, Cadence Design Systems, San Jose, CA., 1994.

[10] Ian Getreu, Analogy Inc., Interview January 1995.

[11] Resve Saleh, Shyh-Jye Jou, A. Richard Newton, *Mixed-Mode Simulation and Analog Multi-Level Simulation*, Kluwer Academic Publishers,

Boston MA, 1994.

[12] Analogy Documentation, *Introduction to the Saber Simulator*, Analogy Inc., 1992.

[13] David R. Coelho, *The VHDL Handbook*, Kluwer Academic Publishers, Boston, 1989.

[14] Donald E. Thomas and Philip R. Moorby, *The Verilog Hardware Description Language*, Kluwer Academic Publishers, Boston, 1991.

[15] Home page of Dolphin Integration: `http://www.dolphin.fr/faq.html#abcd`, 1997.

[16] Dolphin Integration, ABCD Reference Manual, 1997.

[17] Cadence, *SpectreHDL Reference Manual 4.3.2*, Cadence Design Systems 1994.

[18] Open Verilog International, *Verilog-A Language Reference Manual - Analog Extensions to Verilog HDL Version 1.0*, IEEE 1996.

[19] VHDL-AMS, *IEEE Standard 1076.1 Language Reference Manual (Draft)*, 1998.

[20] Ira Miller, Motorola Inc - Open Verilog International Chairman, personal contact, April 1997.

[21] HDL-A is a registered trademark of ANACAD Computer Systems GmbH.

[22] Anacad, *HDL-A Training Course Notes*, Anacad Inc., 1994.

[23] Analogy, *MAST Modeling Class Notes*, Analogy Inc. 1991.

[24] MATLAB is a registered trademark of MathWorks, Inc.

[25] SPW is a registered trademark of Cadence Design Systems, Inc.

[26] David J. Roulston, *Bipolar Semiconductor Devices*, McGraw-Hill, New York, 1990.

[27] MicroSim Corporation, *The Design Center - Application Notes Manual Version 5.4*, MicroSim Corporation, Irvine CA, 1993.

[28] Paul A. Duran, "Behavioral Modeling of a DC-to-DC Converter", Master's Thesis, Massachusetts Institute of Technology, Department of Electrical Engineering and Computer Science, Cambridge MA, 1993.

3
Methodology

Understanding different categories of simulation tools and how they fit into a design methodology can be very confusing. This chapter elaborates on three basic IC design methodologies: bottom-up design, top-down design, and hierarchical design; it discusses how the simulation tools can be used with respect to each of these methodologies. The chapter concludes with a suggested analog behavioral modeling methodology for functional blocks in an IC system.

3.1 Design Methodologies

There is no set design methodology that is guaranteed to work all the time. Every design is unique, every design is similar, and every designer is different. With so many variables it is important for today's designer to be knowledgable of all the design tools and be open to new methods. To complicate matters, there are many simulation tools to choose from. A particular tool may be too much; another tool may not be adequate. Another issue, in regard to design tools and methodology, is that sometimes the tools need to be adapted to fit a particular design methodology. It is the proverbial chicken and egg - as methodologies evolve the tools have to adapt, but sometimes new methodologies cannot be explored until the tools exist.

This section elaborates on basic design methodologies and suggests various categories of design tools that can be used for these methodologies.

3.1.1 Bottom-Up Design

Bottom-up design is traversing the modeling continuum from more detail to less detail. See Figure 3.1 for a graphical representation of bottom-up design.

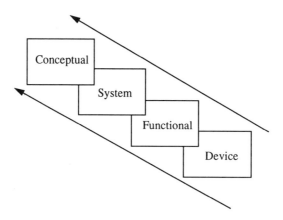

Figure 3.1 Bottom-up design: moving up the abstraction hierarchy.

More specifically, the bottom-up design flow can be explained by the following steps:

1) *Device-level blocks are first designed and simulated using a device-level simulator (e.g. SPICE).*

   Bottom-up design begins by thinking of the individual blocks of a system. Device process characteristics such as: thresholds, betas, current driving capability, and saturation are the first items to consider. Geometric layout requirements are also important since this directly affects cost. Noise and bandwidth considerations are also important when designing from the bottom-up.

Design Methodologies

2) *The crucial interfaces between the functional blocks of system are verified through SPICE or behavioral simulation.*

   The functional blocks are then put together and the system design usually takes place after the individual blocks have been designed. This results in the tweaking of the individual blocks until they interact with the system as expected. If drastic design changes are made, re-design of all the individual blocks may be the only solution to correcting the problems.

3) *Compatibility between the IC topology/architecture is verified with the external application through behavioral or HLD simulation (e.g. Matlab, SPW programs, etc.).*

   Breadboards are more commonly used to verify that the IC meets the requirements of the external system. Sometimes the silicon is used to determine compatibility with the external system which can be very costly if many IC revisions have to be made.

Advantages of the bottom-up design approach include: designer familiarity with the simulation tools; each piece of the system can be quickly designed at the device-level.

A disadvantage of this approach is that system performance cannot be verified until all the blocks have been designed. This can lead to major design changes late in the design-cycle. Another disadvantage is that the architecture and topology cannot be optimized for best system performance. Modest-sized designs, however, using this methodology have proven to be successful. [1]

3.1.2 Top-Down Design

Top-down design is traversing the modeling continuum from less detail to more detail. See Figure 3.2 for a graphical representation of top-down design.

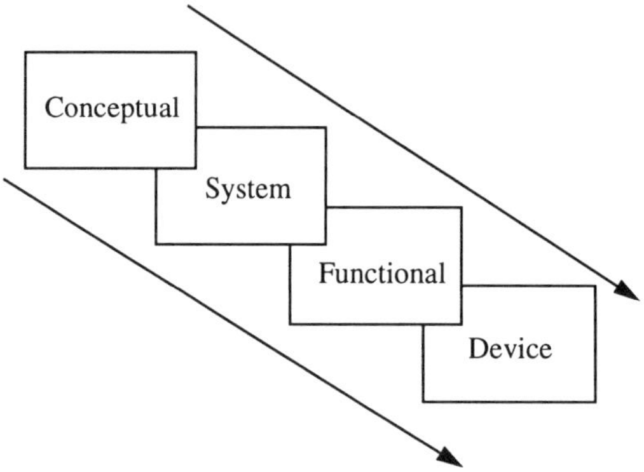

Figure 3.2 Top-down design.

The following is the procedure for the top-down design flow:

1) The design begins with simulation of the specified system topology/ architecture. The models can usually be implemented in a high-level design tool (e.g. SPW, Matlab) and then later transferred to behavioral code once the algorithms have been developed.

Considering architectural and topological issues of the system before any detailed design takes place is important for optimizing system performance. Once an architecture or topology has been selected, the functional system blocks can then be specified.

2) The functional blocks of the entire system are defined, modeled, and verified through behavioral modeling, macromodeling, and simulation. The functional models are then used to verify the interfaces of the system.

The entire system can then be simulated with all the individual functional blocks together. The blocks are defined at the functional-level not considering the device configuration.

3) Device-level design takes place once all the blocks have been specified. The individual blocks are usually simulated using SPICE simulations.

Once the block definition has been defined to the point where minimal change is necessary, device-level design is then performed.

One advantage of the top-down design approach is that the IC can be optimized for system performance before any detailed design is performed. This will save time by keeping design iteration minimal later in the design-cycle. Another advantage of top-down design is that larger and more complex systems can be designed more easily. Connecting the system specification to the block definition is also a very important advantage.

A disadvantage of this approach is the greater learning curve for becoming efficient with the top-level design tools. Another disadvantage is that realization of the design in silicon is not considered. For example, a device process may pose power consumption, cost, die size, etc., constraints that are not considered early in the design-cycle. [1][2]

3.1.3 Hierarchical Design

Hierarchical design is traversing the modeling continuum by designing top-down and verifying bottom-up. See Figure 3.3 for a graphical representation of hierarchical design.

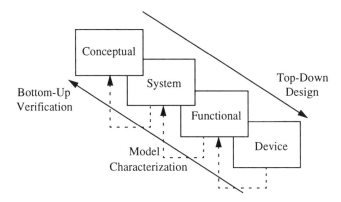

Figure 3.3 Hierarchical design methodology.

The following is the procedure for the hierarchical design flow:

1) Top-down design procedure.

2) Bottom-up design procedure with model characterization.

   2.1) The functional block models are characterized by extracting the functional parameters from the SPICE simulations.

   2.2) The system is verified for functionality and connectivity through the simulation of all the characterized functional blocks.

   2.3) The system topology/architecture model can be characterized by the functional block-level system model for key parameters. The top-level can be modified to incorporate second-order effects that were determined to be important from the lower-level simulations.

   2.4) The system can then be verified for compatibility between the IC and the external application. Also, process considerations (i.e. output drive capability, heat, noise, e.t.c) can be made since the underlying device information is contained within the system model from the model characterization.

   2.5) The high-level description can then be characterized from the system simulations.

   2.6) The high-level description can be used to further verify the system with the external application if the system definition requires too much simulation time at lower-levels of abstraction.

Hierarchical design combines both the bottom-up and top-down approaches. The hierarchical approach begins with analyzing the topological and architectural system issues. Unlike the top-down design methodology, process considerations are made as well, so that design issues at the silicon level can be addressed early in the system definition. Requirements for output drive capability, noise, matching, heat, die size, and junction parasitics should be considered, when possible, at the same time that top-level decisions are made. The device process is usually the limiting factor when trying to meet the system requirements. [3]

High-level behavioral or algorithmic models can be used to select the best system architecture and topology suited for the respective application. The necessary detail of the model is dependent on what is being analyzed.

Design Methodologies

The following example demonstrates the hierarchical design methodology for a basic gain function:

The top-level model can initially be described as a constant times an input voltage. See Equation 3.1 for the high-level gain function relationship and Figure 3.4 for a schematic representation.

$$V_{out} = V_{in} \times K \qquad \text{(Eq 3.1)}$$

where K is a constant. This model could be implemented algorithmically or directly in a behavioral model.

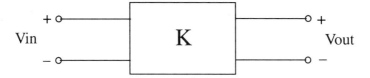

Figure 3.4 Schematic representation of the high level-gain function model.

Once the topology and architecture have been defined, the individual blocks that make up the system must be specified. Each block has requirements that must be met so that all blocks interact appropriately with the overall system. In the gain block example, this is the realization of the gain block being a voltage divider with impedances Z1 and Z2. See Equation 3.2 for the functional gain block relationship and Figure 3.5 for a schematic representation.

$$V_{out} = \frac{Z_1}{(Z_1 + Z_2)} \times V_{in} \qquad \text{(Eq 3.2)}$$

where Z1 and Z2 are constants.

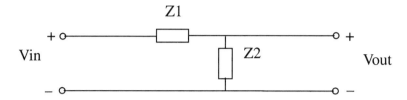

Figure 3.5 Schematic representation of the functional gain block model.

Through behavioral macromodeling each block can be modeled and then implemented together to form a model of the entire IC to verify system functionality and connectivity. If necessary, more model detail can be added to those blocks sensitive to process parameters (temperature, parasitics, drive limits, etc.).

Once all the blocks have been defined each block can be designed at the device-level based on the specific requirements for each block. In the gain block example, this could mean implementing the impedances as *jfet* devices. See Equation 3.3 and Figure 3.6 for the device implementation of the gain block.

$$V_{out} = \frac{Z(L, W, T, t)_1}{Z(L, W, T, t)_1 + Z(L, W, T, t)_2} \times V_{in} \quad \text{(Eq 3.3)}$$

where L, W, T, and t are the *jfet* impedance parameters length, width, temperature, and time respectively.

Design Methodologies

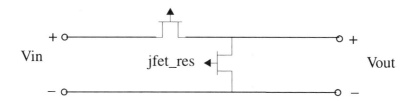

Figure 3.6 Schematic representation of the device-level gain block model.

As each block is designed at the device-level, it can be verified through SPICE simulation. Through multi-level simulation, a single block in the system can be simulated at the device-level, with all the other blocks modeled at the functional-level (this is not possible using SPICE; it is possible through analog behavioral modeling and simulation). In the gain block example, this implies modeling Z1 at the functional-level and Z2 at the device-level. See Figure 3.7 for a representation of this multi-level simulation schematic.

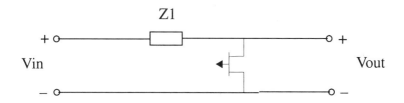

Figure 3.7 Multi-level simulation schematic.

Multi-level simulation is used to verify that each block simulated at the device-level, will interact with the other blocks at the system-level. This is an example of modeling and simulation with overlapping abstraction levels of the modeling continuum.

SPICE simulations can then be used to extract the important functional parameters that will characterize the higher-level analog behavioral models.

For example, SPICE simulation results can be used to extrapolate bandwidth, or temperature effects that could play an important role system functionality. Characterizing the behavioral models will make it possible to retain most of the model precision with respect to the device-level models while simulating the entire system at the functional-level behaviorally. In the gain block example, this implies that each impedance can be characterized to incorporate bandwidth and other important second-order information. Specifically, Z1 and Z2 can be implemented as functions of their respective lengths, widths, temperature, time, etc.

Once all the blocks have been designed and verified, the functional-level system model can be used to characterize the top-level system model. In the gain block example, that implies extrapolating second-order information from the functional-level simulation results to characterize the top-level model. Thus, we have Equation 3.4 where K is no longer a constant.

$$V_{out} = V_{in} \times K(L, W, T, t)_{12} \quad \text{(Eq 3.4)}$$

The top-level model can then be used to verify compatibility between the IC system and the respective application. The fictitious loop has been closed once the design has been verified through the hierarchy of abstraction (the modeling continuum) back to the top-level. Hence, top-down design with bottom-up verification has been demonstrated.

This methodology allows for the entire system to be optimized at all levels, thereby verifying the design from the device blocks up to the IC system. Designing through the hierarchy allows for the obvious errors to be worked out early in the design-cycle. This will lead to minimal changes once the device-level blocks have been designed. The idea is to iterate the design as much as possible early in the design-cycle before fabrication. The slightest errors that are discovered can save another trip to the mask shop and debugging time in the laboratory. Through bottom-up characterization the connection between the top-level model and the actual devices has been maintained (i.e. traceability to silicon), thereby retaining important design information that can be simulated more efficiently at higher-levels of abstraction.

Disadvantages of this approach include: the initial time needed to learn the simulation tools to implement the models, the time to create good characterization data of the underlying device process so that trade-offs between process variation and the system can be made early in the design-cycle, and the time invested in designing quality models that contain the necessary amount of information needed to abstract the important behavior.

3.2 Analog Behavioral Modeling Methodology

Writing or implementing analog behavioral models in an AHDL is not difficult. Determining what to model and what level of abstraction the model should encompass is usually the most difficult task. Basically, if the behavior can be described by an equation or boolean expression, a model can be created. There is no exact method to create a model; however, it is important to remember to be organized, especially when writing many models for a large design. See Figure 3.8 for a flowchart of a suggested analog behavioral modeling methodology for modeling functional blocks.

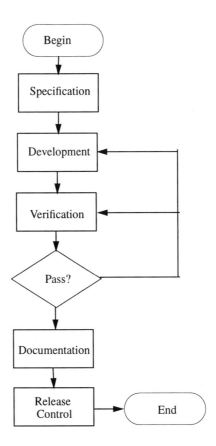

Figure 3.8 Analog Behavioral Modeling Flow.

3.2.1 Specification

Model specification is where the model definition is conceived. The actual algorithm may be sometimes more efficiently derived in a spreadsheet, MATHCAD or MATLAB. This will allow the developer to focus on the model equations without having to worry about implementing them in a simulator. Initially, a few basic questions need to answered before developing the specification:

Analog Behavioral Modeling Methodology

What level of detail will the model incorporate?
What are the model limitations?
What interaction will this model have with other models?

Next, many aspects of the model that will evolve into the model specification need to be considered. The following table shows a few model characteristics and examples that specify a model.

Model Characteristic	Examples
input/output pins	p(plus pin), m (minus pin)
pin types	electrical, mechanical, digital
model parameters	delays, initial conditions
model constraints	voltage thresholds
logic levels	5.0, 3.3 volt logic
error/warning reporting	"WARNING (VCC - GND) is out of range"
model behavior	Vout = $V_A + V_B$

Lastly, a virtual test fixture should be specified. This test fixture will be used in an acceptance test for model accuracy and precision.

3.2.2 Development

The model development occurs once all the requirements have been established. The first step in the development process is to develop the algorithm. This can be done on paper or through algorithmic modeling and simulation. The second step is to use model diagrams or pseudo-code to assist in understanding how the different functions can be implemented in the AHDL. There are many methods to implement a function in an AHDL, so it is best to study the model so that it can be implemented efficiently. The final step is to write the code.

3.2.3 Verification

Model verification is the essential step of making sense of the simulation results. If the models are simple functions they can be verified by direct observation of the simulation results. However, it is usually best practice to verify the model with SPICE simulation once the blocks have been designed. Sometimes actual silicon or a breadboard can also be used to verify the models. In a mixed-signal model, the digital components can be verified through digital simulation. Once confidence is gained in the model through verification, it can be used in the simulation of larger systems.

3.2.4 Documentation

Once the model has been developed and verified it is also best practice, as with all programming/coding, to formalize documentation of the model. Models written months ago may be very difficult to understand if the model is not well documented. If there are many modelers, it is best to establish a standard for documentation so that anyone who is part of the design team can understand the model with little effort. It is also important to communicate the details of the model so that the model can be reused by other design teams. In short, good documentation will reduce technology redundancy and facilitate technology sharing.

3.2.5 Release Control

Releasing a model is usually the step in modeling that most model developers neglect. Throwing a model over the fence once it has been completed can very dangerous. During model development, best practice suggests that a configuration management scheme be put into place. This will allow the design team to keep track of updates to the models, older revisions, and design changes. Another aspect of release control is to create a repository for completed models that is made accessible to all design team members as well as to other design teams in other business units. This will help facilitate technology sharing.

REFERENCES

[1] Paul A. Duran, "Behavioral Modeling of a DC-to-DC Converter", Master's Thesis, Massachusetts Institute of Technology, Department of Electrical Engineering and Computer Science, Cambridge MA, 1993.

[2] H. Alan Mantooth and Mike Fiegenbaum, *Modeling with an Analog Hardware Description Language*, Kluwer Academic Publishers, Boston, 1995.

[3] Paul Duran and Mike Vicker, "Hierarchical Verification of a DC-to-DC Converter through Analog Behavioral Modeling", Power Conversion and Intelligent Motion, pp. 58-64, April 1996.

4
Basic Building Blocks

This Chapter begins with a mini tutorial on MAST. It also demonstrates the implementation and verification of basic analog behavioral modeling building blocks.

4.1 MAST Mini-Tutorial

The intent of this section is not to teach a complete course in MAST, but to give an overview of the fundamentals so that the reader can understand the model examples which are written in MAST.

A model template is used to create a MAST model. See Example 4.1 for an example of a model template.

```
template <name><pins> = <template parameters>
<global declarations>
{ #begin model body
<local declarations>
parameters{
          <parameters assignments>
          }
```

```
<Netlist Section>
when{
     <state (digital) expressions>
     }
values{
       <values equations>
       }
control_section{
                <simulator control statements>
                 }
equations{
          <node current input/output equations>
          }
} #end model body
```

Example 4.1 Mast model template

A MAST template can be created using any text editor (e.g. vi). The template file name should be in the form: <filename>.sin. There are seven basic types of sections within a template that may or may not be used depending on the model. The following are short descriptions and examples of each type of section:

template header - where the template name, pin names, pin types, and parameters are designated

```
template resistor plus minus = rvalue, temp
electrical plus, minus
state logic_4 digout
```

global declarations -   designated after the template header; default values of the parameters can be assigned in this section

```
number rvalue= 1k, temp = 72
```

MAST Mini-Tutorial

local declarations - designated after the first ellipses; this section contains the declaration of local variables

```
number constant
val p power
```

parameter section - defines variables whose values do not change with simulation; parameterizing a model allows for the model to become generic so that the model can be used for many cases

```
parameters{
    constant = temp/25
         }
```

netlist section - netlist statements can be made throughout the template; however, for consistency it is good practice to define these statements between the *parameters* and *when* sections; MAST templates support netlist hierarchy (instantiating of a model within the model) - this subject is beyond the scope of this book

```
r.rout v1 gnd = 10k
```

when section - defines digital behavior and state variables; in this section: digital events can be scheduled, events on state or digital pins can be monitored; state variable values change with simulation

```
when(~dc_init){
     if(power > 10) {
        schedule_event(time,digout,14_0)
        message("warning!")
        message("power = %",power)
                 }
         else schedule_event(time,digout,14_1)
               }
```

values section - supports the *when* and *equations* sections by defining the relationships for the dependent variables; this section is evaluated throughout the simulation

```
values{
    vpm = v(plus) - v(minus)
    ir = vpm/(constant*rvalue)
    power = vpm*ir
    }
```

control_section - sets the simulator convergence parameters for models that require specific model points to be recognized by the simulator (generally for nonlinear models); the specific points are defined in the *parameters* section

```
parameters{
            sv = [(-1meg,1u),(01u),(1meg,0)
            nv = [(-1,2),(1,0)]
            }
    control_section{
                    sample_points(vin,sv)
                    newton_step(vin,nv)
                    }
```

equations section - establishes the relationship for *the across* and *through* variables at the input/output pins; each *through* (i.e. current) path of the model is defined in this section [1]

```
equations section{
                i(plus->minus) += ir
                }
```

The reader should not focus on the language syntax because languages will always be changing - model content is more important.

The next section of this book is dedicated to model examples. The first examples will initially be simple and become increasing complex. The more complex examples will contain a corresponding test circuit and simulation results.

Electrical Sources

All schematics were drawn in DesignStar. The netlists were created from the schematics. All simulations were performed in the Saber Simulator.

4.2 Electrical Sources

The following examples demonstrate how to implement the basic models for voltage and current sources.

4.2.1 DC Current Source

The DC current source is one of the easiest models to write. The output is defined to be:

$$io = cval$$

where *cval* is the DC current value.

The model code is straightforward. See Example 4.2 for the DC current source model code. Note that the output current is assigned directly.

```
template csource plus minus = cval
electrical plus, minus
number i cval
{
equations {
          is: v(plus)-v(minus) = cval
          }
}
```

Example 4.2 Current source model code.

4.2.2 DC Voltage Source

The voltage source is defined by the following equation:

$$vo = vval$$

where *vval* is the DC voltage value.

The model code is also straight forward. See Example 4.3 for the DC voltage source model code. The input pin is defined to be *plus* and the output pin is defined to be *minus*. *vval* is a parameter to the model that will be defined by the user when this model is instantiated. The *equations* section can be interpreted as follows: define the current (*is*) through the model so that the voltage across the model is equal the specified DC voltage (*vval*).

```
template vsource plus minus = vval
electrical plus, minus
number v vval
{
var is
equations {
          i(plus->minus) += is
          is: v(plus)-v(minus) = vval
          }
}
```

Example 4.3 DC voltage source model code.

4.2.3 Voltage Controlled Voltage Source

The voltage controlled voltage source is a little more complicated. The output voltage is governed by the following equation:

$$vo = vin*c$$

where *c* is a constant.

The model code is similar to the voltage source code. However, a *values* section is used to establish the output voltage relationship. See Example 4.4 for the voltage controlled voltage source model code.

```
template vcvs vinp vinm vop vom = c
electrical vinp, vinm, vop, vom
number nu constant
{
number v vo
var is
values {
        vin = v(vinp)-v(vinm)
        vo  = c*vin
        }
equations {
          i(vop->vom) += is
          is: v(vop - v(vom) = vo
```

}
}

Example 4.4 Voltage controlled voltage source model code.

4.2.4 Current Controlled Current Source

The output of the current controlled current source (*cccs*) is governed by the equation:

$$iout = in*c$$

where *c* is constant.

This model code is more complex than the voltage controlled voltage source. The model is in basically the same format in that there are two inputs and two outputs. However, this model is special in that an externally defined model *short* is used. See Example 4.5 for the current controlled current source model code. The model *short* is instantiated in the netlist section of the template:

```
short.1 cinp cinm
```

This instantiating of *short* will connect the *cccs* model between the pins *cinp* and *cinm*. See Example 4.6 for the *short* model code.

The input current in the *cccs* model is determined in the *values* section and the output is established in the *equations* section.

```
template cccs cinp cinm cop com = c
electrical cinp, cinm, cop, com
number nu c
short.1 cinp cinm
{
number i iin
number i iout
values{
     iin = i(short.1)
     iout = iin
     }
equations {
        is: v(cop)-v(com) = c*iout
        }
```

}

Example 4.5 Current controlled current source model code.

The short model is simply the DC voltage source model with *vval* set to zero.

```
template short plus minus
electrical plus, minus
{
var is
equations {
        i(plus->minus) += is
        is: v(plus)-v(minus) = 0
        }
}
```

Example 4.6 Short model code.

4.2.5 Exponential Sinusoidal Voltage Source

The exponential sinusoid can be described by the following equation:

$$vo = vos + amp \cdot e^{-a \cdot (time - td)} \cdot \sin(w \cdot (time - td) + ph)$$

where vos = offset voltage
 amp = amplitude of the sinewave function
 a = slope of the exponential function
 w = frequency in radians
 $time$ = independent variable
 td = time delay
 ph = phase.

The code implementation is the most complex of all the sources described thus far. See Example 4.7 for the exponential sinusoidal voltage source model code. [2][3]The user defined parameters: *a, f, amp, ph, vos, td,* and *a. w,* and *rad* are calculated in the *parameters* section, since these values do not change with simulation. The output voltage is calculated in the *values* section; if time is less than the specified time delay, the output will be equal to the offset voltage; otherwise, the output will equal the exponential sinusoid. *limexp* is a special function used for the exponential. This function will prevent mathematical overflow by limiting its value. *step_size* is a simulator function that controls the analog time step of the simulator; the time step should be at least one twentieth of the sinusoidal frequency. The output of the model is finally established in the *equations* section.

```
template expsin p m = f,amp,ph,vos,td,a
electrical p,m
number f,amp=1,ph=0,vos=0,td=0,a=500
{
number w,rad,step
var i iout
val v vo
parameters{
            w = 2*3.14*f
            rad = ph*3.14/180
            step = 1/(f*20)
            }
values{
        if (time <= td) {
            vo = vos
                        }
        else {
            vo = vos+amp*limexp(-a*(time-
                td))*sin(w*(time-td)+ph)
            }
        step_size = step
      }
equations{
            i(p->m) += iout
            iout: v(p)-v(m) = vo
            }
}
```

Example 4.7 Exponential sinusoidal voltage source model code.

The test circuit is simply the exponential sinusoidal source connected to an output node and ground. See Figure 4.1 for the test schematic.

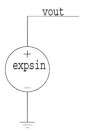

Figure 4.1 Exponential sinusoidal voltage source test circuit.

A 10 millisecond transient simulation was performed on the test circuit. The parameter values were set as follows: *vos* =2, *amp* = 1, *f* = 1k, *td* = 1m, and *a* = 500. The test results can be found in Figure 4.2. The graph reveals a sinusoidal signal that is proportional to a decaying exponential. A time delay of 1 millisecond was specified; this is consistent with the test results.

Voltage Arithmetic

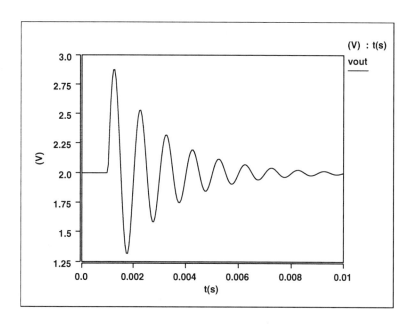

Figure 4.2 Exponential sinusoidal voltage source test results.

4.3 Voltage Arithmetic

The following examples demonstrate how to implement models involving voltage arithmetic: addition, subtraction, multiplication, division, differentiation, and integration.

4.3.1 Voltage Addition, Subtraction, Multiplication, and Division

Voltage addition, subtraction, multiplication and division can be implemented in the same model using the following algorithm:

$$\text{if operator} = 1 \text{ (addition)}; \quad vo = va + vb$$
$$\text{if operator} = 2 \text{ (subtraction)}; \quad vo = va - vb$$

if operator = 3 (multiplication); $vo = va * vb$

if operator = 4 (division); $vo = va/vb$

The model code is similar to a voltage source code. See Example 4.8 for the *arithmetic* model code. The calculation is performed in the *values* section of the model template. *Operator* is a parameter to the model which will be user defined.

```
template arithmetic v1 v2 gnd vout = operator
electrical v1,v2,gnd,vout
number operator
{
var i iout
val v va,vb,vo
values{
        va = v(v1) - v(gnd)
        vb = v(v2) - v(gnd)
        if       (operator == 1) vo = va+vb
        else if (operator == 2) vo = va-vb
        else if (operator == 3) vo = va*vb
        else if (operator == 4) vo = va/vb
        }
equations{
        i(vout->gnd) += iout
        iout: v(vout)-v(gnd) = vo
          }
}
```

Example 4.8 Model code for arithmetic template.

The test circuit used to verify the model consists of a 2 volt peak-to-peak sinusoid (with a 2 volt DC offset), and a 2 volt constant voltage source. These voltage sources are the inputs to two *arithmetic* templates. See Figure 4.3 for the schematic of the test circuit. The upper *arithmetic* model is designated to perform multiplication; the lower is designated to perform addition through the user defined *operator* parameter.

Voltage Arithmetic

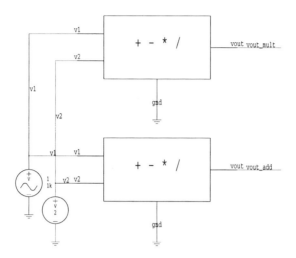

Figure 4.3 Test circuit for the arithmetic model.

The results are shown in Figure 4.4. The lower graph shows the inputs v1 and v2; the middle graph shows the result of the addition of the two inputs; the upper graph shows the multiplication of the two inputs.

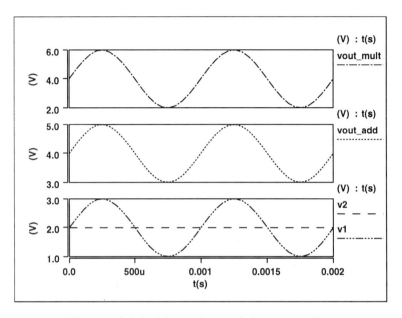

Figure 4.4 Arithmetic model test results.

Any math equation can be used in the *arithmetic* model. For example, a squarer, absolute value, trigonometric function, etc. can be applied to the input.

4.3.2 Voltage Differentiation and Integration

Another mathematical example is voltage differentiation. The algorithm is simply:

$$vo = \frac{d}{dt}(vin)$$

The model code is again straightforward. See Example 4.9 for the model code for voltage differentiation. Note that the differentiation function is performed in the *equations* section.

```
template diff vin gnd vout
electrical vin,gnd,vout
```

Voltage Arithmetic

```
{
var i iout
val v va,vo
values{
        va = v(vin)-v(gnd)
        vo = v(vout)-v(gnd)
        }
equations{
        i(vout->gnd) += iout
        iout:vo = d_by_dt(va)
          }
}
```

Example 4.9 Differentiator model code.

The test circuit for this model consists of a piece-wise linear voltage source that is defined as a linear increasing ramp from 0 to 2.8 volts in 2 miroseconds and then a decreasing ramp to zero volts, also in 2 microseconds. The piece-wise linear voltage source is connected to the input of the differentiation model. See Figure 4.5 for the schematic of the differentiation test circuit.

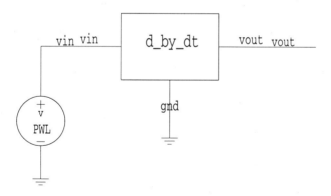

Figure 4.5 Differentiation test circuit.

The test results for the differentiation test circuit can be found in Figure 4.6. The upper graph is the input; the lower graph is the output, which is the derivative of the input.

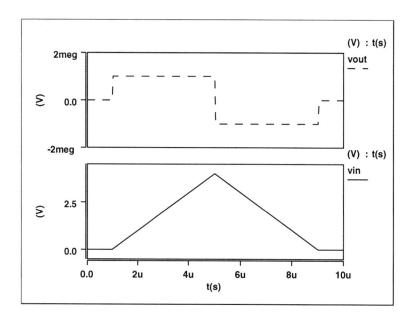

Figure 4.6 Differentiation test results.

Integration can be performed in the same way. However, the integral must be found indirectly by transforming the equation in terms of a derivative.

$$vo = \int (vin)dt$$

$$\frac{d}{dt}vo = vin$$

See Example 4.10 for the voltage integration model code.

```
template int vin gnd vout
```

```
electrical vin,gnd,vout
{
var i iout
val v va,vo
values{
    va = v(vin)-v(gnd)
     vo = v(vout)-v(gnd)
     }
equations{
     i(vout->gnd) += iout
     iout: d_by_dt(vo) = va
        }
}
```

Example 4.10 Voltage integration model code.

4.4 Electrical Primitives

The following section will show model examples of the basic electrical elements: resistor, capacitor, inductor, diode, and transistor.

4.4.1 Resistor

The model behavior for a resistor is described by the equation:

$$iout = \frac{vres}{res}$$

where *vres* is voltage across the resistor and *res* is the value of the resistance.

The model code is straightforward as well. However, to make things more interesting a calculation for power dissipation in the resistor has been added to the model code. See Example 4.11 for the resistor model code. The *val*, *power*, can be accessed after simulation and displayed along with the voltages and currents in a transient simulation.[4] This feature can be very useful when monitoring the total power dissipation in a circuit.

```
template res p m = rval
```

```
electrical p, m
number rval
{
val v vres
val i ires
val p power
values{
 vres = v(p) - v(m)
 ires =  vres/res
 power = vres * ires
      }
equations {
          i(p->m) += ires
          }
}
```

Example 4.11 Resistor model code.

4.4.2 Capacitor

The capacitor model is very similar to the differentiator model. A parameter has been added to specify the capacitance value as well as the equation for charge in the capacitor. The output current is the derivative of the charge. See Example 4.12 for the capacitor model code.

```
template cap p m = cval
electrical p, m
number cval
{
val q qcap
val v vcap
values{
  vcap = v(p)-v(m)
 qcap = vcap*cval
      }
equations {
          i(p->m) += d_by_dt(qcap)
          }
}
```

Example 4.12 Capacitor model code.

Electrical Primitives

4.4.3 Inductor

The inductor model is also similar to the differentiator model. The output voltage is the derivative of the current. See Example 4.13 for the inductor model code.

```
template ind p m = ival
electrical p, m
number ival
{
var ind
equations {
  i(p->m) +=  ind
  ind: v(p)-v(m) = d_by_dt(ival*ind)
        }
}
```

Example 4.13 Inductor model code.

4.4.4 Ideal Diode

The main equation that governs diode current is:

$$id = is(e^{(vin)/(vt)} - 1)$$

This equation can be directly implemented in the model. See Example 4.14 for the ideal diode model code. The two main parameters which characterize the diode are the saturation current *is* and the thermal voltage *vt*.[5]

```
template idiode p m = is,vt
electrical p,m
number is=1E-14,vt=26m
{
val v vd,vs
val i id
values{
      vd = v(p) - v(m)
      vs = vd/vt
      id = is*(limexp(vs)-1)
```

```
        }
equations{
    i(p->m) += id
        }
}
```

Example 4.14 Ideal diode model code.

The test circuit for the ideal diode model is simply a DC voltage source connected to the positive side of the diode. See Figure 4.7 for the test circuit schematic.

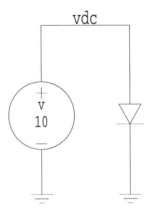

Figure 4.7 Ideal diode test circuit.

The test results of a DC sweep are shown in Figure 4.8. The results reveal the expected ideal diode current as a function of the diode voltage.

Electrical Primitives

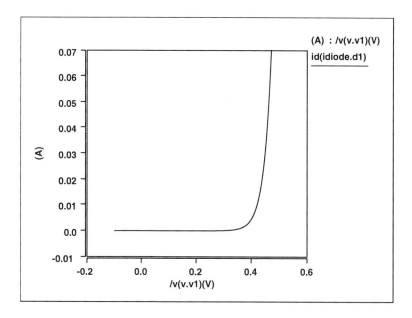

Figure 4.8 Ideal diode model test results.

4.4.5 Ideal Transistor

Transistor behavior can be modeled as an ideal switch. The model can be configured so that the switch emulates a npn, pnp, nmos, or pmos transistor. This approach will reduce the amount of simulation time needed to simulate this type of device. In this example, we will implement the n-type (npn/nmos) ideal transistor. This behavior can described by switching an impedance between a large and small value based on the input control signal (i.e. gate voltage):

Input High (Logic 1) implies vout = iout * ron (switch on)

Input Low (Logic 0) implies vout = iout * roff (switch off)

These two conditions can be added together to a single output equation:

$$vo = y \cdot iout \cdot ron + (y - 1) \cdot iout \cdot roff$$

where:

$y = 1$ when switch is *on*

$y = 0$ when switch is *off*

Changing the value of *y* instantaneously could cause discontinuities in the model behavior during simulation. Therefore, when this variable changes, it is good modeling practice to ramp the value of *y* up and down in a linear fashion. Using the standard equation for a line, Equation 4.1 was derived:

$$y = \frac{x}{m} \cdot (ynew - yold) + yold \qquad \text{(Eq 4.1)}$$

This equation descibes both, ramping up and down:

$ynew = 1$, $yold = 0$ implies ramp-up

$$y = \frac{x}{m}$$

$ynew = 0$, $yold = 1$ implies ramp-down

$$y = -\frac{x}{m} + 1$$

See Example 4.15 for the ideal transistor model code.[3] In the model, *t1-time* is substituted for *x* which translates the x-access to the origin when an event occurs on the *control* input and *dt* is substituted for *m* which is the change in time between transitions; *ron* is the on resistance and *roff* is the off resistance of the switch; *dt*, *ron*, and *roff* are user defined parameters. Checking for digital events on the *control* pin is performed in the *when* section of the template. *yold* and *ynew* are also defined in the *when* section. The equation for *y* is in the *values* section and the voltage across the switch is established in the

Electrical Primitives

equations section.

```
template itran ctrl p m = ron, roff, dt
state logic_4 ctrl
electrical p,m
number ron=.001,roff=100meg,dt=1p
{
var i iout
val v vout
val nu y
state time t1 = -1
state nu ynew,yold
state logic_4 old_ctrl=l4_1
when(dc_done){
    old_ctrl == ctrl
   if(ctrl == l4_x | ctrl == l4_z | ctrl == l4_1){
        ynew = 1
                                                    }
    else ynew = 0
            }
when(event_on(ctrl)){
    if(ctrl ~= l4_x & ctrl ~= l4_z & ctrl ~=
       old_ctrl & time >= t1){
       old_ctrl=ctrl
       yold=y
       if(ctrl == l4_0){
          ynew = 0
                     }
       else ynew = 1
       if(time_domain){
         schedule_next_time(time)
         t1 = time + dt
         if(dt~=0) schedule_next_time(t1)
                      }
                                                    }
                   }
values{
     if(time>=t1) y = ynew
     else         y = (t1-time)/dt*(ynew-yold
                     ) + yold
     vout = v(p) - v(m)
```

81

```
        }
equations{
    i(p->m)  +=iout
    iout:vout = y*iout*ron + (1-y)*iout*roff
        }
}
```

Example 4.15 Ideal transistor model code.

The test circuit for the ideal transistor model includes: a clock source set to a frequency of 1 kiloHertz, a 10 volt positive rail voltage source, and a 1 kilo ohm load resistor. See Figure 4.9 for the test circuit schematic.

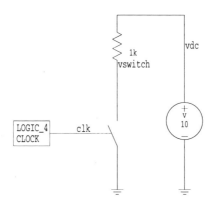

Figure 4.9 Ideal transistor test circuit.

A 5 millisecond transient analysis was performed on the test circuit. The test results for the ideal transistor test can be found in Figure 4.10. The lower graph reveals that the switched node, *vswitch*, transitions between the rail voltage and ground according to the clock signal, *clk* (input to *control*). The clock signal is shown in the upper graph. This is the expected behavior for an ideal npn or nmos transistor. The behavior of the switch can be changed to emulate a pnp or pmos transistor by inverting the *control* input signal inside the model.

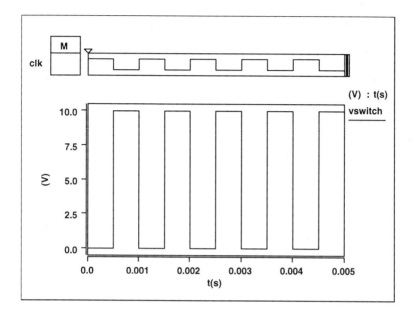

Figure 4.10 Ideal transistor test results.

REFERENCES

[1] H. Alan Mantooth and Mike Fiegenbaum, *Modeling With An Analog Hardware Description Language*, Kluwer Academic Publishers, Boston, 1995

[2] Muhammad H. Rashid, *SPICE For Circuits And Electronics Using PSpice*, Prentice Hall, Englewood Cliffs, New Jersey, 1990.

[3] Analogy Inc., *MAST Modeling Class Notes*, Published with permission from Analogy Inc., 1991.

[4] Analogy Inc., Patent Number 4,868,770.

[5] Paul R. Gray and Robert G. Meyer, *Analysis of Design of Analog Integrated Circuits -Third Edition*, John Wiley & Sons, Inc.1993.

5

More Building Blocks

This Chapter defines a variety of more complex models which can be found in IC systems. These models are implemented as the ideal case for simplicity of demonstration. Each model is designed to be a foundation upon which higher-order models can be written. Specifically, analog, digital, mixed signal, and mixed-system models are presented in this chapter.

5.1 Analog Models

The intent of this section is to demonstrate how analog building blocks can be implemented in an AHDL. Specifically, the examples in this section include: ideal transformer, peak detector, sample-and-hold, Schmitt trigger, voltage-to-frequency converter, and frequency-to-voltage converter analog blocks.

5.1.1 Ideal Transformer

The transformer is an analog block commonly found in power systems. The current and voltage equations which describe the ideal transformer are:

$$vs = vp \cdot N$$

$$ip = -is \cdot N$$

$$N = \frac{n2}{n1}$$

where vs = secondary voltage

vp = primary voltage

is = secondary current

ip = primary current

N = turns ratio

n1 = number of primary windings

n2 = number of secondary windings.

These equations are based on three basic assumptions: flux leakage is zero, magnetizing current is zero, and the transformer works under DC and AC conditions; hence, the windings are perfectly coupled. These equations also assume that the windings traverse the core in the same direction. The code implementation is straightforward. [1] See Example 5.1 for the ideal transformer code. The template pins include: *pp* primary positive input, *pm* primary minimum input, *sp* secondary positive input, and *sm* secondary minus input. The user defined parameters are *n1* and *n2*, the number of windings for the primary and secondary coils, respectively. The secondary voltage, *vout*, is calculated in the *values* section. The secondary current, *is*, is set by the output load and reflected back to the primary; this assignment is also made in the *equations* section, *i(pp->pm)*.

```
template tf pp pm sp sm = n1,n2
electrical pp,pm,sp,sm
number n1=1,n2=1
{
number n
var i is
val v vout,vin
```

Analog Models

```
parameters{
  n = n2/n1
      }
values{
  vin = v(pp) - v(pm)
  vout = vin*n
      }
equations{
  i(sp->sm) += is
  iout: v(sp) - v(sm) = vout
  i(pp->pm) += -is*n
      }
}
```

Example 5.1 Ideal transformer model code.

The test circuit used to verify the model consists of a sinusoidal input voltage source set at 2 volts peak-to-peak and a secondary load of 1 ohm. *Short* models are inserted on each side of the transformer to demonstrate that the current relationship holds by plotting the current through each *short*. See Figure 5.1 for the schematic of the test circuit.

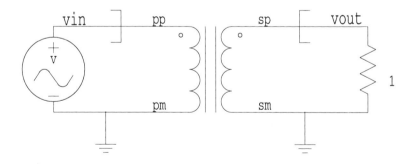

Figure 5.1 Ideal transformer test circuit.

The following corresponding test circuit netlist was generated:

87

```
short.shorts m:b p:vout
short.shortp m:a p:vin
r.r1 m:0 p:vout = rnom=1
v.v1 m:0 p:vin = ac=(mag=1,phase=0),//
tran=(sin=(va=2,f=1k))
tf.tf1 pm:0 pp:a sm:0 sp:b = n2=1, n1=2.
```

Nodes *a* and *b* are not shown in the figure for simplicity. // denotes line continuation.

A DC and a 10 millisecond transient analyses were performed on the test circuit:

```
dc
tr (siglist / /short.*/i,tend 5m,tstep 1u).
```

The above code shows what would be typed in at the command line of the Saber simulator window to invoke simulation.

The test results are shown in Figure 5.2. The primary voltage *vin* and the secondary voltage *vout* are displayed in the upper graph. The primary current *i(shortp)* and the secondary current *i(shorts)* are displayed in the lower graph. The results reveal that the model is performing as expected.

Analog Models

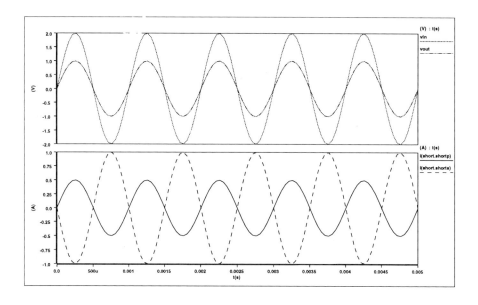

Figure 5.2 Ideal transformer test results.

5.1.2 Peak Detector

Peak detectors are analog blocks that can be found in test/measurement instrumentation and in amplitude modulation systems. The ideal peak detector can be described by the following algorithm:

if vin >= vout (previous state)   vout = vin

else   vout = vout (previous state).

In an actual peak detector a capacitor will store the peak voltage level and will discharge at a rate proportional to a resistor/capacitor time constant. [2] The model code is straightforward for the ideal peak detector. See Example 5.2 for the peak detector model code. The template pins include: *in* input, *out* output, and *gnd* ground. An intermediate state variable, *vout,* is defined with units of voltage. During DC initialization *vout* is set to zero and during the transient domain (i.e. time domain), the condition for the input voltage, *vin,* is compared against the previous state, *vout*. State variables can only be set in the *when* section of the model template. The output current, *iout*, is specified in the *equations* section.

```
template peakd in out gnd
electrical in,out,gnd
{
val v vin
var i iout
state v vout=0
when(dc_init) {
  vout = 0
            }
when(time_domain) {
  if(vin >= vout) vout = vin
                }
values{
  vin = v(in) - v(gnd)
     }
equations{
  i(out->gnd) += iout
  iout: v(out)-v(gnd) = vout
        }

}
```

Example 5.2 Peak detector model code.

See Figure 5.3 for the test circuit schematic. This test circuit consists of a piecewise linear input voltage source connected to the input of the peak detector.

Analog Models

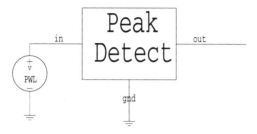

Figure 5.3 Peak detector test circuit.

The following corresponding test circuit netlist was generated:

```
v.v1 m:0 p:vin = //
tran=(pwl=[0,0,1m,2,2m,1,3m,3,4m,2,5m,4,6m,0])
peakd.p1 out:vout in:vin gnd:0.
```

A DC and a 6 milli-second transient analyses were performed on the test circuit:

```
dc
tr (tend 6m,tstep.1n).
```

A 0.1n time step was used because more resolution was needed to keep the state *vout* updated in the peak detector model with respect to the input voltage each time the previous peak was exceeded. The results are shown in Figure 5.4. The input *vin* is a sawtooth waveform with various peaks. The output *vout* is always equal to the maximum of the input voltage.

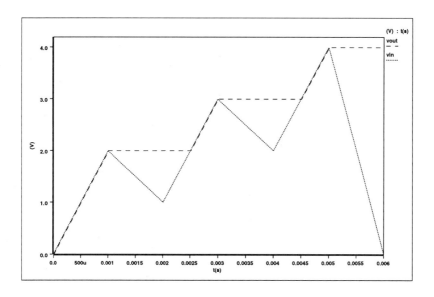

Figure 5.4 Peak detector test results.

5.1.3 Sample-and-Hold

The sample-and-hold (S/H) function is used in analog-to-digital conversion (ADC) systems; it is more commonly found in successive approximation ADC systems. Other applications include: analog demultiplexing in data distribution, analog delay lines, and electronic musical keyboards. The S/H function captures analog data and holds it for processing. [2] This function can be described by the following algorithm:

$$\begin{aligned}&\text{if} \quad \text{clock} = \text{logic } 1 \quad vout = vin\\&\qquad\qquad\qquad\qquad\qquad vlast = vin\\&\text{else if } \text{clock} = \text{logic } 0 \quad vout = vlast\end{aligned}$$

where vout = output voltage

Analog Models

vin = input voltage

vlast = held output voltage.

This algorithm describes ideal S/H behavior. The code implementation is straightforward. See Example 5.3 for the S/H model code. This model has 4 template pins: *in* input, *out* output, *gnd* ground, and *clock* the sample clock. This clock signal will carry the S/H command during each clock cycle. The output *out* is an analog state. Implementation of an analog state will allow the simulator to converge more easily and faster, since the value on this pin is not introduced into the system matrix. If the output were specified as it is in a voltage source (specifying an output current and voltage in the *equations* section) the value of *vout* would have to be ramped-up or ramped-down to the next value. As mentioned previously, discontinuities involving across and through variables will cause problems with simulation convergence and in some cases will lead to erroneous results.

For simplicity, the S/H model has been implemented using an analog state. If this model were to be used to drive an external load, a z-domain-to-analog converter model would have to be used to covert the state representation of the output voltage into a respective current/voltage node. Another option is to ramp the output in the model up and down accordingly, as was done in the ideal transistor example in Chapter 4. *vlast* and *vout* are state variables that will be used to set *out*, the state output pin. In the *when* section, the clock is tested for logic 1. If the *clock* is a logic 1, *vout* is set to *vin*. Otherwise, it will be set to the last value of vin (*vlast*) before the negative clock edge. *vlast* is set every time the *clock* is a logic 1. The state output (*out*) will be scheduled only if the last value scheduled does not equal the next determined output value (i.e. *out* does not equal *vlast*). Hence, the amount of events scheduled via the event-queue will be reduced by scheduling events only when necessary; thereby reducing the simulation time. Finally, *vin* is determined in the *values* section.

```
template sh in out gnd clock
electrical in,gnd
state v out
state logic_4 clock
{
val v vin
```

```
state v vout=0,vlast=0
when(time_domain) {
   if(clock == 14_1) {
        vout = vin
        vlast = vin
                    }
   else vout = vlast
   if(out ~= vlast) schedule_event(time,out,vout)
                    }
values{
  vin  =  v(in) - v(gnd)
     }
}
```

Example 5.3 Sample-and-hold model code.

The test circuit used to verify the S/H model consists of a piece-wise linear input voltage source set to a ramp input waveform from 0 to 4 volts in 5 milliseconds and then back down to 0 volts, also in 5 milliseconds. A logic clock source is used as the input to the S/H command pin *clock*. The clock is set to a frequency of 1 kHz and duty-cycle of 50 percent. See Figure 5.5 for the test circuit schematic.

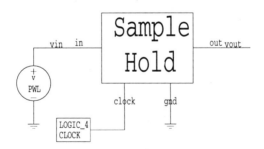

Figure 5.5 Sample-and-hold test circuit.

The following corresponding test circuit netlist was generated:

Analog Models

```
v.v1 m:0 p:vin = tran=(pwl=[0,0,5m,4,10m,0])
clock_14.ck1 clock:clock = freq=1k, duty=.5
sh.sh1 out:vout clock:clock in:vin gnd:0.
```

A DC and a 10 millisecond transient analyses were performed on the test circuit:

```
dc
tr (tend 10m,tstep.1n).
```

The results are shown in Figure 5.6. The logic clock waveform is displayed in the upper graph. The input *vin* and output *vout* are displayed in the lower graph. The input is sampled on every positive clock edge and held until the next cycle. The results reveal the expected performance of an ideal S/H circuit.

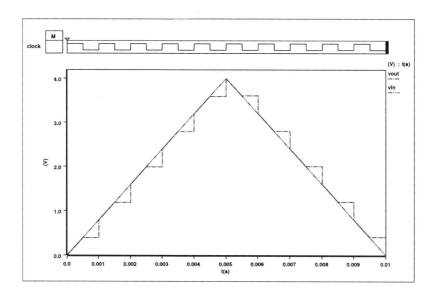

Figure 5.6 Sample-and-hold test results.

5.1.4 Non-inverting Schmitt Trigger

The Schmitt trigger function is used to reduce noise in comparators and is used to build oscillators. The Schmitt trigger uses positive feedback to saturate the amplifier upon which it is built, thereby allowing only two stable states, positive and negative saturation voltages. The special quality of a Schmitt trigger is that it exhibits hysteresis. [2] The following algorithm describes a non-inverting Schmitt trigger:

$$\text{if} \quad vin > vt\_h \quad vout = vs\_h$$
$$\text{else if} \quad vin < vt\_l \quad vout = vs\_l$$

where vt_h = high trigger voltage
vt_l = low trigger voltage
vs_h = high saturation voltage
vs_l = low saturation voltage
vin = input voltage
vout = output voltage.

See Example 5.4 for the Schmitt trigger model code. The code implementation will again use an analog state as the output *vout*. There are only three pins in the model: *in* input, *out* output, and *gnd* ground. The user defined parameters are: *vs_h* saturation high voltage, *vs_l* saturation low voltage, *vt_h* trigger high voltage, and *vt_l* trigger low voltage. To produce the hysteresis effect two sets of threshold flags need to be used, *before1*, *before2*, *after1*, and *after2* in the *when* section. These flags are set when the high and low trigger voltages are crossed by the input voltage; the corresponding voltage state is scheduled on the output during each of these events. At DC the output is simply set to zero in the *when* section.

```
template st in out gnd = vs_h, vs_l, vt_h, vt_l
electrical in, gnd
state v out
number vs_h, vs_l, vt_h, vt_l
```

Analog Models

```
{
val v    vin
state nu before1,before2,after1,after2
when(dc_init) {
schedule_event(time,out,0)
            }
when(threshold(vin,vt_h,before1,after1)) {
   if((before1 == -1) & (after1 == 1)) {
        schedule_event(time,out,vs_h)
                                        }
                                          }
when(threshold(vin,vt_l,before2,after2)) {
   if((before2 == 1) & (after2 == -1)) {
        schedule_event(time,out,vs_l)
                                        }
                                          }
values{
   vin  = v(in) - v(gnd)
        }
}
```

Example 5.4 Non-inverting Schmitt trigger model code.

The test circuit used to verify the Schmitt trigger model consists of a piece-wise linear input voltage source set to ramp-up from 0 to 4 volts in 2 microseconds and ramp-down from 4 to 0 volts, also, in 2 microseconds. The user defined parameters are set as follows: $vs\_h = 5$, $vs\_l = 0$, $vt\_h = 3$, and $vt\_l = 0$. See Figure 5.7 for the test circuit schematic.

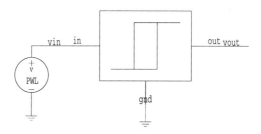

Figure 5.7 Non-inverting Schmitt trigger test circuit.

The following corresponding test circuit netlist was generated:

```
v.v1 m:0 p:vin = tran=(pwl=[0,0,2u,4,4u,0])
st.st1 out:vout in:vin gnd:0 = \\
vs_h=5,vt_l=2,vt_h=3,vs_l=0.
```

A DC and a 5 microsecond transient analyses were performed on the test circuit:

```
dc
tr (tend 5u,tstep 1n).
```

The results are shown in Figure 5.8. The upper graph displays the input and output voltage. The output *vout* goes to *vs_h* as soon as the input crosses the 3 volt *high* trigger level and the output goes to *vs_l* as soon as the input crosses the 2 volt *low* trigger voltage. The lower graph displays the output voltage as a function of the input voltage; this characteristic is known as a hysteresis curve. The results reveal the expected performance of an ideal non-inverting Schmitt trigger.

Analog Models

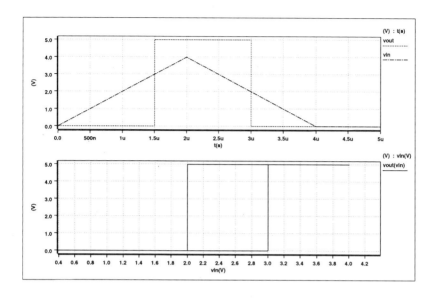

Figure 5.8 Non-inverting Schmitt trigger test results.

5.1.5 Voltage-to-Frequency Converter

The voltage-to-frequency converter (VFC) is a specific type of voltage controlled oscillator (VCO) used in high performance applications. VFCs can be found in digital volt meters, phase-locked loops, tone decoders, and PWM controllers. The ideal VFC function is described by the following algorithm:

$$\text{if (vin = 0)} \quad fout = fbase$$
$$\text{else} \quad fout = k \cdot vin$$

$$vout = a \cdot \sin(2 \cdot \pi \cdot fout \cdot t)$$

where fout = output frequency
 fbase = base or free running frequency

99

k = proportionality constant (sensitivity)
vout = output voltage
vin = input voltage.

This algorithm describes behavior only for sine wave output. A complete VFC will produce triangular and square output waveforms. However, for demonstration purposes, we will only analyze the sinusoidal case. See Example 5.5 for the VFC model code. The template has 3 pins: *in* input, *out* output, and *gnd* ground. The user defined parameters include: *f* free running frequency, *amp* amplitude, and *k* constant. As with the exponential sinusoid example the step size for the analog simulation is calculated to be one-twentieth of the free running frequency in the *parameters* section. The output voltage is set in the *values* section. And finally, the output current and voltage are set in the *equations* section.

```
template vf in out gnd   = f,amp,k
electrical in,out,gnd
number  f,amp=1,k=1
{
number step
var i iout
val v vin,vo
val nu w
parameters{
          step = 1/(f*20)
          }
values{
       vin = v(in) - v(gnd)
       w = 2*3.14*f*k
       if(vin == 0)   vo = amp*sin(w*time)
       else           vo = amp*sin(w*vin*time)
       step_size = step
       }
equations{
          i(out->gnd) += iout
          iout: v(out)-v(gnd) = vo
          }
}
```

Example 5.5 Voltage-to-Frequency converter model code.

Analog Models

The test circuit used to verify the VFC model consists of a piece-wise linear input voltage source set to 0 volts between 0 and 5 milliseconds and ramp-up to 2 volts between 5 and 5.01 milliseconds. This setting should initially establish the free running frequency and then multiply the input frequency by 2 at 5.01 milliseconds. The user defined parameters were set as follows: *k=1*, *amp=2*, *f=1k*. See Figure 5.9 for the test schematic.

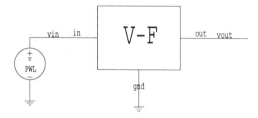

Figure 5.9 Voltage-to-Frequency converter test circuit.

The following corresponding test circuit netlist was generated:

    v.v1 m:0 p:vin = tran=(pwl=[0,0,5m,0,5.01m,2])
    vf.vf1 out:vout in:vin gnd:0 = k=1, amp=1, f=1k.

A DC and a 10 millisecond transient analyses were performed on the test circuit:

    dc
    tr (tend 10m,tstep 1u).

The results are shown in Figure 5.10. The output waveform is displayed in the upper graph and the input waveform in the lower graph. Initially, the output is a sinusoid oscillating at the free running frequency 1 kilohertz. As the input transitions from 0 to 2 volts, the output frequency doubles as expected.

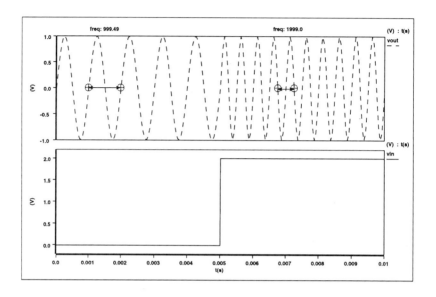

Figure 5.10 Voltage-to-Frequency converter test results.

5.1.6 Frequency-to-Voltage Converter

The frequency-to-voltage converter (FVC) is the inverse function of the VFC block. FVCs are used together with VFCs to couple analog information in an isolated form. Opto-isolators, fiber optic links, pulse transformers, and RF links are examples of isolators that can be inserted between a VFC and a FVC for coupling. The ideal FVC can be described by the following algorithm:

$$fd = fmax - fmin$$

$$fin = \frac{1}{T}$$

$$vout = k \cdot \frac{fin - fmin}{fd}$$

where fmin = minimum input frequency

Analog Models

fmax = maximum input frequency

fd = frequency range

fin = input frequency

T = period of input waveform

vout = output voltage

k = proportionality constant (sensitivity).

The output voltage is established as a ratio of the difference between the input frequency and the minimum frequency to the frequency range. See Example 5.6 for the FVC model code. This model works only for sinusoidal inputs. The template pins include: *in* input, *out* output, *gnd* ground. The user defined parameters are: *vdc* DC offset voltage, *fmin* minimum input frequency, *fmax* maximum input frequency, and *k* constant.

The difficulty in this model is calculating the period of the input sinusoid. This is done by setting one of two time flags each time the input crosses the DC threshold *vdc*. At this time, half of the period, *td* is calculated by calculating the time elapsed since the previous threshold crossing (e.g. *td* = *time - t1*, where *t1* was set during the previous crossing). The full period is calculated by multiplying *td* by 2 in the *values* section. During a DC analysis *vout* is set to 0 volts by setting the period equal to the minimum frequency. The flags *t1* and *t2* are also initialized during DC. The input frequency *fin*, and the output voltage *vout* are calculated in the *values* section. The output voltage and current are established in the *equations* section.

```
template fv in out gnd   = vdc,fmin,fmax,k
electrical in,out,gnd
number vdc=0,fmin,fmax,k=1
{
number fd
state nu before,after,td,t1,t2
var i iout
val v vin,vo,fin
parameters{
          fd = fmax - fmin
          }
when(dc_init) {
```

103

```
            td = 1/(2*fmin)
            t1 = 0
            t2 = 0
                }
        when(threshold(vin,vdc,before,after)) {
            if  ((before == -1) & (after == 1)) {
                t1 = time
                td = time - t2
                                                            }
            else if((before == 1) & (after == -1)) {
                t2 = time
                td = time - t1
                                                            }
                                                                }
        values{
                vin = v(in) - v(gnd)
                fin = 1/(2*td)
                vo = (fin-fmin)/fd
            }
        equations{
                i(out->gnd) += iout
                iout: v(out)-v(gnd) = vo
                }
    }
```

Example 5.6 Frequency-to-Voltage converter model code.

The test circuit used to verify the FVC model consists of the VFC test circuit which is used to establish the input stimulus for the FVC. The same settings were used for the piecewise-liner voltage and the VFC model. The user defined parameters for the FVC model were set as follows: *fmin = 1k, fmax = 3k, k = 1, vdc = 0*. These parameters were set on the basis that the output from the VFC test circuit was used to drive the FVC block (i.e. 1 kHz input signal should produce 0 volts and the 2 kHz signal should produce 0.5 volts since it is mid-range between the maximum and minimum input frequency). See Figure 5.11 for the test schematic.

Analog Models

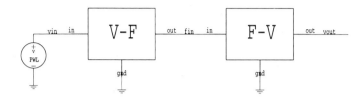

Figure 5.11 Frequency-to-Voltage converter test circuit.

The following corresponding test circuit netlist was generated:

```
fv.fv1 out:vout in:fin gnd:0 = fmin=1k, k=1,\\
       vdc=0, fmax=3k
v.v1 m:0 p:vin = tran=(pwl=[0,0,5m,0,5.01m,2])
vf.vf1 out:fin in:vin gnd:0 = k=1, amp=1, f=1k.
```

A DC and a 10 millisecond transient analyses were performed on the test circuit:

```
dc
tr (tend 10m,tstep 1u).
```

The results are shown in Figure 5.12. The output voltage is displayed in the upper graph and the input voltage is displayed in the lower graph. The input is 0 volts when the input frequency is the minimum (1 kHz.) and ramps-up to the expected .5 volts when the input frequency is mid-range (2 kHz.).

105

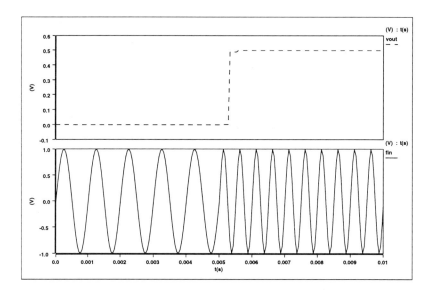

Figure 5.12 Frequency-to-voltage converter test results.

5.2 Digital Models

This section demonstrates how digital behavior can be modeled in an AHDL. The examples include: an AND gate, a multiplexer, and a D-latch. A single test circuit was used to verify all three gates together.

5.2.7 AND Gate

The AND gate can be described by the following truth table:

in1	in2	out
0	0	0
0	1	0

Digital Models

in1	in2	out
1	0	0
1	1	1
x	1 or 0	x
1 or 0	x	x

where in1, in2 = data inputs

out = data output

note: x implies an unknown state.

The AND function is easily implemented in MAST. See Example 5.7 for the AND gate model code. The template pins include: *in1* 1st input, *in2* 2nd input, and *out* output. For simplicity, the only user defined parameter is *td* propagation delay; this is the time from when there is an event on an input to the time a change on the output occurs. The actual template is less complex than those for analog blocks since the inputs and outputs are digital states. The Boolean function is performed in a *when* section by checking for events on the inputs. The output is set through the *schedule_event* command.

```
template and in1 in2 out = td
number td=0
state logic_4 in1,in2,out
{
when(event_on(in1) | event_on(in2)) {
        if  (in1 == l4_1 & in2 == l4_1) {
           schedule_event(time+td,out,l4_1)
                                          }
          else if(in1 == l4_x | in2 == l4_x){
             schedule_event(time+td,out,l4_x)
                                             }
          else schedule_event(time+td,out,l4_0)
                                          }
}
```

Example 5.7 AND gate model code.

5.2.8 Multiplexer

The multiplexer function can be described by the following truth table:

a0	a1	out
0	0	in0
0	1	in1
1	0	in2
1	1	in3

where in0, in1, in2, in3 = data inputs

out = data output

a0, a1 = select inputs.

In this example, x-states (unknown states) have been omitted for simplicity. See Example 5.8 for the multiplexer model code. The template pins include: *a0* and *a1* select inputs, *in0 - in3* data inputs, and *out* output. The user defined parameter is *td* propagation delay.

```
template mux a0 a1 in0 in1 in2 in3 out = td
number td=0
state logic_4 a0,a1,in0,in1,in2,in3,out
{
when(event_on(a0) | event_on(a1)) {
        if(a0 == 14_0 & a1 == 14_0){
           schedule_event(time+td,out,in0)
                                          }
        else if(a0 == 14_0 & a1 == 14_1){
           schedule_event(time+td,out,in1)
                                          }
        else if(a0 == 14_1 & a1 == 14_0) {
           schedule_event(time+td,out,in2)
                                          }
        else if(a0 == 14_1 & a1 == 14_1) {
           schedule_event(time+td,out,in3)
```

Digital Models

```
                                            }
         else schedule_event(time+td,out,14_x)
                                }
}
```

Example 5.8 Multiplexer model code.

5.2.9 D-Latch

The D-latch function can be described by the following table:

d	q	qb
0	0	1
1	1	0

where d = data input

q = output

qb = output inverse.

The data will only be clocked in on the positive clock edge. See Example 5.9 for the D-latch model code. Again, for simplicity, x-states have been omitted. The template pins include: *d* data input, *clk* input clock, *q* and *qb* outputs. The only user defined parameter is *td* propagation delay. For efficiency, the model will schedule only events on the output when the output does not equal the previous input (*d* does not equal *q*). Both *q* and *qb* are set if the Boolean condition is met. During a DC analysis, the outputs are simply set to zero.

```
template dlch d clk q qb = td
number td=0
state logic_4 d,clk,q,qb
{
when(dc_init) {
   schedule_event(time+td,q,14_0)
   schedule_event(time+td,qb,14_0)
              }
```

```
        when(event_on(clk) & d ~= q) {
            if(clk == 14_1 & d == 14_1) {
                schedule_event(time+td,q,14_1)
                schedule_event(time+td,qb,14_0)
                                                }
            if(clk == 14_1 & d == 14_0) {
                schedule_event(time+td,q,14_0)
                schedule_event(time+td,qb,14_1)
                                                }
                                                }
}
```

Example 5.9 D-latch model code.

See Figure 5.13 for the test circuit schematic that was used to verify all three digital models. The circuit consists of 4 fixed-bit sources used as the data inputs to the multiplexer, 1 fixed-bit source used as one of the inputs to the AND gate, 2 programmable bit-stream sources used as the select inputs to the multiplexer, and a clock source set to 1kHz with a duty-cycle of 50 percent used as the input to the D-latch. The multiplexer output is used as the input to the AND gate and the AND gate output is used as the input to the D-latch. See Table 5.1 data input bits and Table 5.2 for data select bit-stream.

Table 5.1 Data Input bits

in0	in1	in2	in3
1	0	1	0

Table 5.2 Data select bit-stream

time	a0	a1
0ms	0	1
2ms	0	0
4ms	1	1

Digital Models

Table 5.2 Data select bit-stream

time	a0	a1
6ms	1	0

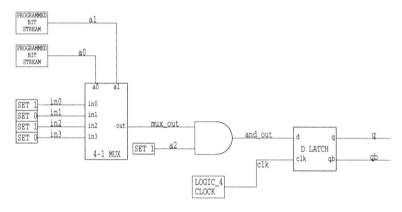

Figure 5.13 Digital test circuit.

The following corresponding test circuit netlist was generated:

```
prbit_14.a1 out:a1 = \\
  bits=[(0,_0),(2m,_1),(4m,_0),(6m,_1)]
prbit_14.a0 out:a0 = bits=[(0,_0),(4m,_1)]
clock_14.clk1 clock:clk = freq=1k, duty=.5
set_14_1.in0 set1:in0
set_14_1.in4 set1:a2
set_14_1.in2 set1:in2
set_14_0.in3 set0:in3
set_14_0.in1 set0:in1
dlch.@"dlch#2" d:and_out qb:qb q:q clk:clk
mux.@"mux#1" out:mux_out a0:a0 a1:a1 in0:in0 \\
            in1:in1 in2:in2 in3:in3
and.@"and#0" out:and_out in1:mux_out in2:a2.
```

A DC and 8 a millisecond transient analyses were performed on the test circuit:

```
dc
tr(tend 8m,tstep 1u).
```

The results are shown in Figure 5.14. All signals are displayed as digital signals. The multiplexer output *mux_out* was, as expected, based on the select inputs *ao* and *a1*. The AND gate performed the correct function and the D-latch latched the data correctly based on the input clock signal *clk*.

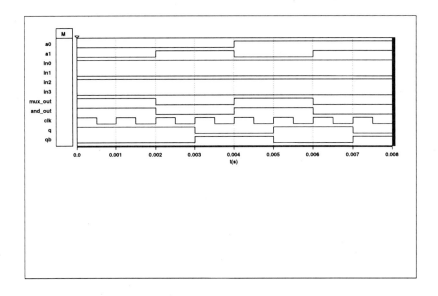

Figure 5.14 Digital test results.

5.3 Mixed Signal Blocks

This section demonstrates how both analog and digital behavior can be modeled together in a single template. This is possible because MAST is a mixed-signal AHDL. Specifically, a voltage comparator, a pulse-width modulator, and an analog-to-digital converter are modeled in this section.

5.3.1 Voltage Comparator

The voltage comparator is a very important block. It can be found in level detectors, Schmitt triggers, switching power supplies, waveform generators, and analog-to-digital converters. [2] The voltage comparator can also be thought of as the simplest form of an analog-to-digital converter. The behavior of a voltage comparator can be described by the following algorithm:

$$\text{if } (vin1 > vin2) \quad vout = logic\ 1$$
$$\text{if } (vin1 < vin2) \quad vout = logic\ 0$$

where $vin1$ = positive input
$vin2$ = negative input.

This algorithm describes ideal comparator behavior. The output of an actual comparator saturates to the positive and negative rail. A real comparator would also exhibit hysteresis effects. Implementation of the voltage comparator in MAST is straightforward. See Example 5.10 for the voltage comparator model code. The input pins are: *in1* positive input, *in2* negative input, and *out* digital output. The user defined parameter is *td*, propagation delay. The difference between the input voltages *vdiff*, is calculated in the *values* section. This difference is then compared against 0 in a *when* section. If *vdiff* is greater than zero the output will be set to a logic 1 after *td* and if *vdiff* is less than zero the output will be set to a logic 0 after *td*. During DC the output is simply set to a logic 0.

```
template cmp in1 in2 out = td
```

```
electrical in1,in2
state logic_4 out
number td=0
{
val v vdiff
state nu before,after
when(dc_init) {
  schedule_event(time,out,14_0)
              }
when(threshold(vdiff,0,before,after)) {
    if((before == -1) & (after == 1)) {
         schedule_event(time+td,out,14_1)
                                       }
    else schedule_event(time+td,out,14_0)
                                        }
values{
  vdiff   = v(in1) - v(in2)
        }
}
```

Example 5.10 Voltage comparator model code.

The test circuit used to verify the voltage comparator model consists of a periodic piecewise-linear positive input voltage sources that was set to ramp-up from 0 to 5 volts in .5 milliseconds and ramp-down to 0 volts in .5 milliseconds. A DC voltage source set to 2.5 volts is used as the negative input to the comparator. The parameter td is set to 0. See Figure 5.15 for the test circuit schematic.

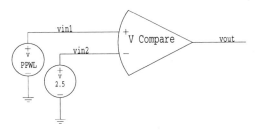

Figure 5.15 Voltage comparator test circuit.

Mixed Signal Blocks

The following corresponding test circuit netlist was generated:

```
v.v1 m:0 p:vin1 = tran=(ppwl=[0,0,.5m,5,1m,0])
v.v2 m:0 p:vin2 = dc=2.5
cmp.c1 out:vout in1:vin1 in2:vin2 = td=0.
```

A DC and a 5 millisecond transient analyses were performed on the test circuit:

```
dc
tr (tend 5m,tstep 1u).
```

The test results are shown in Figure 5.16. The lower graph displays the positive and negative comparator input voltages. The upper graph shows the output toggling to a logic 1 when *vin1* is greater than *vin2* and toggling to a logic 0 when *vin1* is less than *vin2*. This result reveals the expected ideal comparator behavior.

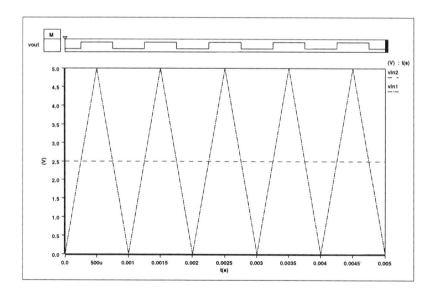

Figure 5.16 Voltage comparator test results.

5.3.2 Pulse-Width Modulator

The pulse-width modulator (PWM) is a common block found in switching power supplies and is also used in signal recording/transmission applications. The PWM output is a pulse train that is modulated by a comparison between input voltage and a sawtooth waveform. [2] The PWM function can be described by the following algorithm:

$$cmpmax = dmax \cdot (sawh - sawl) + sawl$$

$$cmpmin = dmin \cdot (sawh - sawl) + sawl$$

if (vin > cmp_max)   vin = cmp_max
else if (vin < cmp_min)   vin = cmp_min
else   vin = vin

if (vin > vsaw)   vout = logic 1
else   vout = logic 0

where: dmax = maximum duty-cycle
dmin = minimum duty-cycle
cmp_max = maximum input voltage to produce maximum allowed duty-cycle
cmp_min = minimum input voltage to produce minimum allowed duty-cycle
vin = input voltage
vout = output voltage.

This algorithm will generate a square waveform with a duty-cycle that is equal to the percentage of time that the input waveform is greater than the sawtooth for a given cycle. The frequency of the square waveform is, therefore, equal to the frequency of the sawtooth waveform. See Example 5.11 for the PWM model code.

Mixed Signal Blocks

The template pins include: *in* input, *saw* sawtooth input, *gnd* ground, *out* output. The user defined parameters include: *saw_h* high sawtooth voltage, *saw_l* low sawtooth voltage, *d_max* maximum duty-cycle, and *d_min* minimum duty-cycle. The maximum and minimum comparison voltage *cmp_max* and *cmp_min*, respectively, are calculated in the *parameters* section. These voltages will set the limits for the voltage that will be compared to the sawtooth waveform in the a *when* section; the *threshold* command was used to perform the comparison. The input voltage is compared to the limits in the *values* section. If the input is within the limits, it will then be compared to the sawtooth waveform to produce an output waveform with the respective duty-cycle in the *when* section.

```
template pwm in saw gnd out = \\
          saw_h,saw_l,d_min,d_max
electrical in,saw,gnd
state logic_4 out
number saw_h,saw_l,d_min=0,d_max=1
{
val v      cmp_in,vsaw,vin
number     cmp_max,cmp_min
state nu before,after
parameters{
   cmp_max = d_max*(saw_h-saw_l) + saw_l
   cmp_min = d_min*(saw_h-saw_l) + saw_l
         }
when(threshold(vsaw,cmp_in,before,after)) {
     if((before == -1) & (after == 1)) {
         schedule_event(time,out,14_0)
                                         }
     else schedule_event(time,out,14_1)
                                                }
values{
   vin  = v(in) - v(gnd)
   vsaw = v(saw) - v(gnd)
   if(vin >= cmp_max & d_max < 1) \\
     cmp_in = cmp_max
   else if(vin <= cmp_min & d_min > 0)\\
     cmp_in = cmp_min
   else cmp_in = vin
       }
```

117

}

Example 5.11 PWM model code.

The test circuit used to verify the PWM model consists of a piece-wise linear input voltage source set to ramp-up between 0 and 5 volts in 10 milliseconds. A periodic piecewise-linear voltage source is used to create the sawtooth waveform with a maximum voltage of 4 volts, minimum voltage of 2 volts, and a frequency of 1 kHz. The user defined parameters of the pwm are set as follows: $d\_max = .8$, $d\_min = .1$, $saw\_h = 4$, $saw\_l = 2$. The saw_h and saw_l parameters are set based on the sawtooth waveform characteristics. See Figure 5.17 for the test circuit schematic.

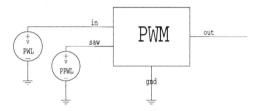

Figure 5.17 PWM test circuit.

The following corresponding test circuit netlist was generated:

```
v.saw1 m:0 p:saw = tran=(ppwl=[0,2,.9m,4,1m,2])
v.pwl1 m:0 p:vin = tran=(pwl=[0,0,10m,5])
pwm.pwm1 out:vout in:vin saw:saw gnd:0 = \\
         d_max=.8, saw_h=4, d_min=.1, saw_l=2.
```

A DC and a 10 millisecond transient analyses were performed on the test

Mixed Signal Blocks

circuit:

```
dc
tr (siglist / /pwm.*/*,tend 10m,tstep 1u).
```

The *siglist / /pwm.*/** command allows the top-level signal to be displayed by */* and those inside the PWM model by */pwm.*/**. The test results are shown in Figure 5.18. The lower graph displays the ramped and sawtooth input voltages. The middle graph displays the internal pwm voltage that is compared to the sawtooth waveform; this voltage is clipped such that a 10 percent minimum and a 80 percent maximum duty-cycle can be achieved. The upper graph displays the digital PWM output which increases from a 10 to 80 percent duty-cycle as the input voltage is ramped through the sawtooth waveform. These results reveal the expected ideal PWM behavior.

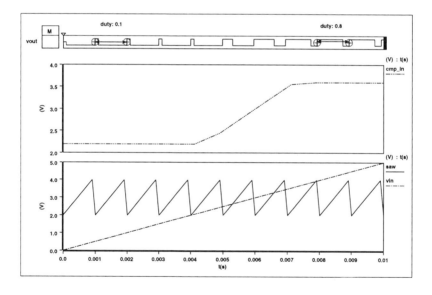

Figure 5.18 PWM test results.

119

5.3.3 Analog-to-Digital Converter

The analog-to-digital converter (ADC) block is a very important mixed-signal building block that converts naturally occurring analog information of voltage, current, charge, temperature, etc., into the more convenient digital form for processing. There are many circuit techniques for analog-to-digital conversion: counter-ramp converters, tracking converters, sigma-delta converters, etc. This example will focus on a type of converter known as a successive-approximation converter (SAC).

The SAC technique will complete an n-bit conversion in n clock periods. In the 1st clock cycle the MSB (2^{nth} place bit) is set by subtracting 2^n from the decimal equivalent of the input voltage and comparing the result to 0. If the result is greater than 0, the *nth*-bit is set to a logic 1 and the remainder is equal to the decimal equivalent of the input voltage minus 2^n. If it is less than 0, the *nth*-bit it is set to a logic 0 and the remainder is equal to the decimal equivalent of the input voltage. On the next clock cycle the 2^{nth-1} place bit is set by comparing the remainder minus 2^{n-1} to 0. This will continue until the 2^0 (LSB) place bit is set. [2] The following algorithm describes an 8-bit DAC that uses successive-approximation conversion:

$$lsb = \frac{Vref}{2^8}$$

$$decin = \frac{Vin}{lsb}$$

$$\sum_{n=8}^{1} y_n = \sum_{n=8}^{1} decin - 2^n \cdot x_n - 2^{n-1}$$

$$x_8 = 0$$

such that:

Mixed Signal Blocks

$$y_8 = decin - 2^8 \cdot x_8 - 2^7$$

$$y_7 = decin - 2^8 \cdot x_8 - 2^7 \cdot x_7 - 2^6$$

.
.
.

$$y_1 = decin - 2^8 \cdot x_8 - 2^7 \cdot x_7 - 2^6 \cdot x_6 - 2^5 \cdot x_5 - 2^4 \cdot x_4 - 2^3 \cdot x_3 - 2^2 \cdot x_2 - 2^1 \cdot x_1 - 2^0$$

if ($y_n > 0$) nth bit = logic 1

$$x_{n-1} = 1$$

else nth bit = logic 0

$$x_{n-1} = 0$$

where   lsb = 1 least significant bit decimal equivalent voltage

decin = decimal equivalent input voltage

$y_n = 2^{nth}$ place compare voltage

x_{n-1} = remainder flag.

The complete digital output of the ADC will represent an 8-bit binary code equivalent to the analog input. The decimal range of an 8-bit ADC is from 0 to 255. In this example, we will assume that the DC offset of the input has been subtracted out from the input. Hence, an input voltage of 0 volts will yield a decimal equivalent of 0.

See Example 5.12 for the ADC model code. The input pins include: *in* analog input, *gnd* ground, *sc* sample clock, *eoc* end of conversion bit, *a* through *h* output bits. The user defined parameters are: *vref* analog voltage conversion range, *td* propagation delay, and *teoc* time for end of conversion after last bit has been set. *td* represents the total time to complete all the 2^{nth}

place comparisons. *teoc* is the time from the last 2^{nth} comparison to the time when the output is considered to be valid. Hence, the settling time of the ADC is *td+teoc*.

The ADC will begin conversion when there is an active low event on *sc*; this condition is monitored in a *when* section. The decimal equivalent *pdecin* is then calculated. If the input exceeds 255 the *pdecin* is clamped to 255; if the input is below 0 the *pdecin* is clamped to 0, and if the input is within the 8-bit resolution, *pdecin* will not be changed. After the decimal equivalent has been calculated, the output bits are set based on the 2^{nth} place comparison. Once the last bit has been set, *eoc* is set to a logic 1 and then to a logic 0 after *teoc*.

```
template a2d in gnd a b c d e f g h sc eoc =\\
        vref, td, teoc
electrical in,gnd
state logic_4 a,b,c,d,e,f,g,h,sc,eoc
number td=0,vref,teoc=10u
{
state nu \\
        x0=1,x1=1,x2=1,x3=1,x4=1,x5=1,x6=1,\\
        x7=1,decin,pdecin
state nu y1,y2,y3,y4,y5,y6,y7,y8
val v vin
number LSB = vref/2**8
when(event_on(sc) & sc == 14_0) {
pdecin = vin/LSB
        if (pdecin > 255)      decin = 255
        else if (pdecin <= 0)decin = 0
        else                   decin = pdecin

y8 = decin - 128
x7 = (y8 >= 0)
        if (x7) schedule_event(time+td,h,14_1)
        else    schedule_event(time+td,h,14_0)

y7 = decin-x7*128-64
        x6 = (y7 >= 0)
        if (x6) schedule_event(time+td,g,14_1)
        else    schedule_event(time+td,g,14_0)
```

Mixed Signal Blocks

```
       y6 = decin-x7*128-x6*64-32
            x5 = (y6 >= 0)
            if (x5) schedule_event(time+td,f,14_1)
            else    schedule_event(time+td,f,14_0)

       y5 = decin-x7*128-x6*64-x5*32-16
            x4 = (y5 >= 0)
            if (x4) schedule_event(time+td,e,14_1)
            else    schedule_event(time+td,e,14_0)

       y4 = decin-x7*128-x6*64-x5*32-x4*16-8
            x3 = (y4 >= 0)
            if (x3) schedule_event(time+td,d,14_1)
            else    schedule_event(time+td,d,14_0)

       y3 = decin-x7*128-x6*64-x5*32-x4*16-x3*8-4
            x2 = (y3 >= 0)
            if (x2) schedule_event(time+td,c,14_1)
            else    schedule_event(time+td,c,14_0)

       y2 = decin-x7*128-x6*64-x5*32-x4*16-x3*8-x2*4-2
            x1 = (y2 >= 0)
            if (x1) schedule_event(time+td,b,14_1)
            else    schedule_event(time+td,b,14_0)

       y1 = decin-x7*128-x6*64-x5*32-x4*16- \\
            x3*8-x2*4-x1*2-1
            x0 = (y1 >= 0)
            if (x0) schedule_event(time+td,a,14_1)
            else    schedule_event(time+td,a,14_0)
schedule_event(time+td,eoc,14_1)
schedule_event(time+td+teoc,eoc,14_0)
                   }
values {
vin = v(in) - v(gnd)
       }
}
```

Example 5.12 Analog-to-digital converter model code.

The test circuit used to verify the ADC consists of a piece-wise linear input voltage source that is set to a series of steps between 0 and 2 volts with a .5 volt increment; the time between steps is set to 2 milliseconds. A clock source set to a frequency of 1kHz and a duty-cycle of .5 is used as the sc input to the ADC. The user defined parameters are set as follows: $td = 0$, $teoc = 10u$, and $vref = 2$. The following table reveals the expected output for each stepped input voltage:

vin	decin	hexout
0	0	00
.5	64	40
1	128	80
1.5	192	C0
2	255	FF

The output is represented as a hexadecimal code. A 1 volt input is considered to be half-scale with a hexadecimal output code of 80. See Figure 5.19 for the test circuit schematic.

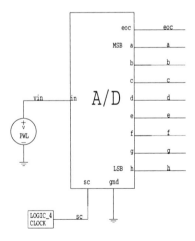

Figure 5.19 Analog-to-digital converter test circuit.

Mixed Signal Blocks

The following corresponding netlist was generated:
```
clock_14.clk1 clock:sc = freq=1k, duty=.5
v.in1 m:0 p:vin = \\
        tran=(pwl=[0,0,2m,0,2.01m,.5,4m,.5,\\
        4.01m,1,6m,1,6.01m,1.5,8m,1.5,8.01m,2])
a2d.ad1 a:a b:b c:c d:d e:e f:f in:vin g:g \\
        h:h sc:sc eoc:eoc gnd:0 = \\
        td=0,teoc=10u, vref=2.
```

A DC and a 10 millisecond transient analyses were performed on the test circuit:
```
dc
tr(tend 10m,tstep 1u).
```

The results are shown in Figure 5.20. The lower graph shows the analog stepped input voltage. The upper graph shows the *eoc*, *sc*, and the output *out_8* represented as an 8-bit output bus with hexadecimal values. The results are as expected. However, there is a small discrepancy with the hexadecimal codes 3F and 7F that have 63 and 127 respective decimal equivalents. This error can be attributed to numerical noise.

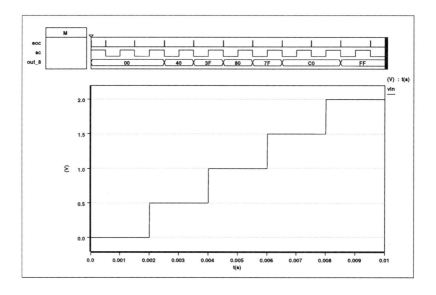

Figure 5.20 Analog-to-digital converter test results.

5.4 Mixed-Signal Interface Models

Mixed-signal interface models are the glue that tie the analog domain to the digital domain and vice-versa. In most mixed-signal simulation environments, the netlister can automatically insert the appropriate interface models depending on the type of conversion (i.e. analog-to-digital or digital-to-analog). Figure 5.21 illustrates the placement on interface models in a circuit that shows a cascade of buffers. The first buffer is represented by a MOS implementation and the second buffer is represented by a digital implementation.

Mixed-Signal Interface Models

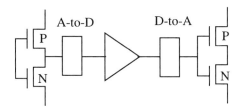

Figure 5.21 Interface model insertion.

If analog output ports are interfaced to digitals input ports, an *A-to-D* interface model is inserted between the analog and digital component via the netlister. If digital output ports are interfaced to analog input ports, a *D-to-A* interface model is inserted between the components via the netlister. Interface models can be defined for any process technology (e.g. TTL, MOS, ECL, etc.). As with any model, various levels of detail can be incorporated into the model. The following examples will show how to build basic interface models for both analog-to-digital and digital-to-analog.

5.4.1 Analog-to-Digital Interface Models

Analog-to-digital interface models convert an analog voltage level to a corresponding digital output (i.e. logic 1 or logic 0). Example 5.13 shows the model code of a basic analog-to-digital interface model.

```
template a2d a m d  = td, vil, vih
electrical a, m
state logic_4 d
number td =0,vil=0.8,vih=2.4
{
state nu before,after
when (dc_init) {
  schedule_event(time,d,l4_0)
 }
# threshold crossing for logic level 0
when(threshold(v(a)-v(m),vil,before,after)) {
   if((after<0) & (driven(d) ~= l4_0)) {
```

```
        schedule_event(time+td,d,14_0)
    }
  }
  # threshold crossing for logic level 1
  when(threshold(v(a)-v(m),vih,before,after)) {
    if((after>0) & (driven(d) ~= 14_1)) {
      schedule_event(time+td,d,14_1)
    }
  }
}
```

Example 5.13 Analog-to-Digital interface model code. [3]

In this model, the parameters are *td* (propagation delay), *vil* (input low voltage), and *vih* (input high voltage). The model determines when the analog input voltage crosses the logic level thresholds (logic 1 and logic 0) and sets the digital output to the appropriate level. The results produced by this model are shown in Figure 5.22.

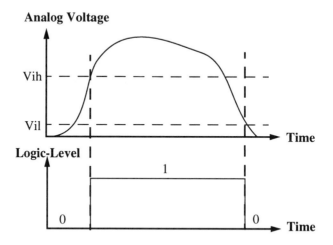

Figure 5.22 Analog-to-Digital interface model results.

Mixed-Signal Interface Models

These results are for an ideal conversion. In reality, there exists an unknown digital state. The upper graph shows the analog input and the lower graph shows the digital output.

5.4.2 Digital-to-Analog Interface Models

Digital-to-analog interface models convert logic levels to analog output voltages. Example 5.14 shows the model code for a basic digital-to-analog interface model.

```
element template d2a d a m = td,vol,voh
electrical a,m
state logic_4 d
number td=0,vol=0.5,voh=4.0
{
var i i
state v vout = vol
when(event_on(d)) {
# logic 0 to analog output conversion
  if(d==l4_0) {
  schedule_event(time+td,vout,vol)
  schedule_next_time(time+td)
}
# logic 1 to analog output conversion
  else if(d==l4_1) {
  schedule_event(time+td,vout,voh)
  schedule_next_time(time+td)
 }
}
equations {
  i(a->m) += i
  i: v(a)-v(m) = vout
 }
}
```

Example 5.14 Digital-to-Analog interface model code. [3]

In this model, the parameters are *td* (propagation delay), *vol* (output low voltage), and *voh* (output high voltage). The model determines the analog output voltage (high or low) based on the digital input logic level. The results produced by this model are shown in Figure 5.23.

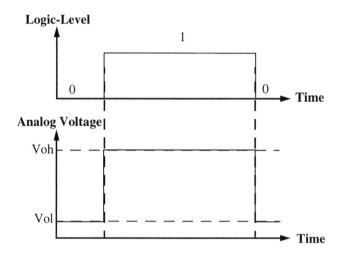

Figure 5.23 Digital-to-Analog interface model results.

These results are also for an ideal conversion. The upper graph shows the digital input and the lower graph shows the analog output.

The Saber simulation environment has model constructs known as Hypermodels™ that act as interface models. The values for the process technology can be changed, if necessary, through model files that define the template parameters for the various technologies (i.e. TTL, CMOS, etc.). Also, various levels of the interface models can be selected (specified in the schematic capture program) if more model detail is necessary.

5.5 Mixed-System Models

This section demonstrates the power of a mixed-system AHDL. Mixed-system AHDL's allow the modeler to design and verify mixed technology systems (e.g. mechanical electrical systems). A DC motor will be modeled in this section.

5.5.3 DC Motor

The DC motor converts electrical energy to mechanical rotational energy. This type of energy conversion is known as an electromagnetic transduction. Hence, a DC motor can be considered an energy transducing system element. [4] There are many configurations for a DC motor: permanent magnet, sperate excitation, shunt excitation, series excitation, and compound excitation. These configurations depend on the technique used to generate the magnetic field that will be cut by the rotor windings. [5] In our example, we will keep things as simple as possible by using a permanent-magnet DC motor. The following is the algorithm which describes ideal DC motor behavior:

$$tgen = kt \cdot iin$$
$$vgen = ke \cdot w$$

where tgen = generated torque

kt = motor toque constant

vgen = generated voltage

ke = motor back-emf constant

iin = input current.

These equations can describe both generator and motor behavior. If an electrical current is applied as an input, the electrical energy will be converter to mechanical rotational energy (i.e. motor transduction) and if a torque is applied as an input the mechanical rotational energy will be converted to electrical energy (i.e. generator transduction). [4] Note that these equations do not include electrical losses due to the resistance/inductance in the windings and mechanical losses due to viscous friction and momentum.

See Example 5.15 for the DC motor model code. [6] The template pins include: *p* electrical positive input, *m* electrical negative input, *shaft* mechanical rotational output, and *rgnd* mechanical rotational ground. The user defined parameters include: *ke* motor back-emf constant, and *kt* motor torque constant. The input terminal voltage *vt* and output angular velocity are calculated in the *values* section. The transduction equations for *vgen* and *tgen* are also in the *values* section. The output torque is specified in the *equations* section, *tq_Nm(shaft->rgnd)*, just as current would be in a current source model. Since *vgen* is dependent on the output torque, its respective terminal current *iin* is also specified in the *equations* section.

```
template mtr p m shaft rgnd = ke,kt
electrical p,m
rotational_vel rgnd,shaft
number ke,kt
{
val t tgen
val v vgen,vt
val w_radps w_rps
var i iin
values {
   vt = v(p) - v(m)
   w_rps = w_radps(shaft)-w_radps(rgnd)
   vgen = ke*w_rps
   tgen = kt*iin
      }
equations {
   i(p->m) += iin
   iin: vt = vgen
   tq_Nm(shaft->rgnd) += tgen
        }
}
```

Example 5.15 DC motor model code.

The test circuit used to verify the DC motor model consists of a piece-

Mixed-System Models

wise liner input current source that is set to ramp-up from 0 to 5 milliamperes in 10 milliseconds. A rotational damper is used as the load with a damping constant set to 1. The rotational damper model produces a damping torque proportional to the angular velocity across the damper:

$$torque_{damping} = d \cdot (velp - velm)$$

where   d = damping constant
   velp = angular velocity at positive input
   velm = angular velocity at minus input.

The reason a load is required is the same reason a load is needed to simulate a current source; the current (torque in this case) needs to flow somewhere. If no load is added, a singular Jacobian error would occur during simulation. Finally, the user defined parameters are set as follows: $kt = 1$, and $ke=1$. See Figure 5.24 for the DC motor test circuit schematic.

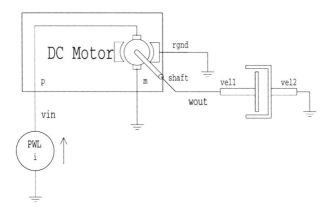

Figure 5.24 DC motor test circuit.

133

The following corresponding test circuit netlist was generated:
```
i.c1 m:vin p:0 = tran=(pwl=[0,0,10m,5m])
damper_w.d1 vel1:wout vel2:0 = d=1
mtr.m1 m:0 p:vin shaft:wout rgnd:0 = kt=1, ke=1.
```

A DC and a 10 millisecond analyses were performed on the test circuit:
```
dc
tr (siglist / /mtr.*/*,tend 10m,tstep 1u).
```

Again, so that the inner variables of the motor template could be displayed in the plot tool, the command `siglist / /mtr.*/*`, was used.

The results are shown in Figure 5.25. The upper graph shows the angular velocity output *wout* of the shaft as a function of the input current *iin* and the lower graph shows the torque output *tgen* as a function of the input current *iin*. The results reveal a linear relationship between the input current and the output torque, and a linear relationship between the input current and the angular velocity which are expected results.

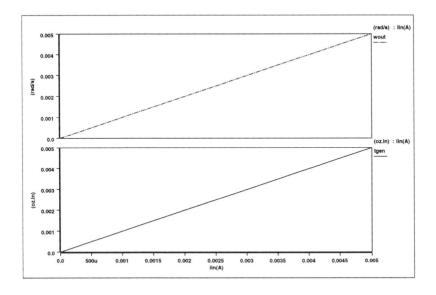

Figure 5.25 DC motor test results.

REFERENCES

[1] John G. Kassakian, Martin F. Schlecht, and George Verghese, *Principles of Power Electronics*, Addison-Wesley Publishing Company, New York, 1991.

[2] Sergio Franco, *Design With Operational Amplifiers and Analog Integrated Circuits, McGraw-Hill Book Company*, New York, 1988.

[3] Analogy Inc., *Guide to Writing Templates Release 3.0*, Published with permission from Analogy Inc., 1992.

[4] Derek Rowell and David N. Wormley, *System Dynamics: An Introduction - Course Notes for Introduction to System Dynamics*, Department of Mechanical Engineering, Massachusetts Institute of Technology, Septem-

ber 1990.

[5] J.F. Lindsay and M.H. Rashid, *Electromechanics and Electrical Machinery*, Prentice-Hall Inc., New Jersey, 1986.

[6] Analogy Inc., *MAST Modeling Class Notes*, Analogy Inc., 1991.

6
IC System Examples

Integrated circuits are evolving into integrated systems. An integrated system may include: microprocessors, firmware, digital signal processors, analog-to-digital and digital-to-analog conversion, sensors, as well as the external system. Today's IC designers are faced with the challenge of obtaining knowledge in the areas of system and IC design, as well as having the design automation knowledge needed to compete large designs. Analog behavioral modeling is one aspect of design automation that will assist designers with system-level issues. The benefits of simulation can be gained both at the component level and at the system-level (i.e. simulating the entire system together hierarchically). At the system-level, designers can: explore architectural and topological trade-offs, optimize the individual components for the best system performance, and understand customer requirements (i.e. think like the customer).

This chapter demonstrates the modeling and simulation of hypothetical IC systems and shows how system-level issues can be investigated. The chapter also explains how to use a hierarchial design methodology to analyze these IC systems. Specifically, a part of a distributed power supply, an audio test system, an automotive ignition system, and a digital communications system are modeled and simulated.

6.1 Distributed Power Supply

In this section, a part of a distributed power supply is modeled and simulated. Specifically, a voltage-mode forward converter and a voltage-mode buck converter will be cascaded to form a part of a distributed power supply model.

6.1.1 System Overview

A basic distributed power supply consists of: an electromagnetic interference input filter stage which prevents back harmonics from escaping the system, a rectifier and peak-detector stage which converts AC to DC, a transformer-isolated converter which generates a *safe* DC voltage that will be distributed to other DC converters that will ultimately drive a variety of loads.

The DC converters can consist of any combination of switching power supply topologies (e.g. buck, boost, and flyback converters). Advantages of distributed power supplies are: each load is individually controlled and tightly regulated, there is isolation and protection from the input stage, and a variety of loads can be efficiently driven. A specific example of a distributed power supply for an application in a personal computer can be found in Figure 6.1. This is a simplified example of a power supply that can drive loads for a CRT display (110 volts, 3 amps), memory (5 volts, 2 amps), and digital logic (3.3 volts, 10 amps). [1]

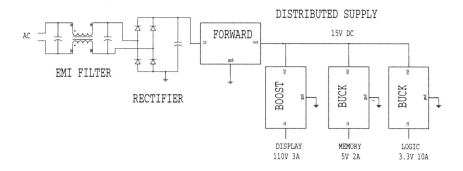

Figure 6.1 Distributed personal computer power supply.

Distributed Power Supply

In this example, we will focus on modeling the forward converter cascaded with a buck converter that will drive 3.3 volt logic at 10 amps (digital logic load). See Figure 6.2 for a schematic of cascaded DC-to-DC switching converters.

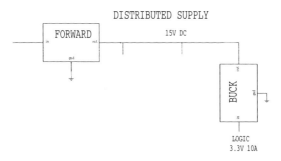

Figure 6.2 Cascaded forward and buck DC-to-DC converters.

A buck (step-down) converter topology produces a lower average output voltage than the DC input. It can easily be recognized by the output low-pass LC filter. See Figure 6.3 for a schematic of a voltage-mode buck converter. The duty-cycle of the switch is regulated by the feedback voltage that is compared to a reference voltage; hence the name, voltage-mode control. The compensation circuit determines the loop stability as well as the converters ability to respond to load and line transients. The PWM usually consists of digital logic, a comparator and an oscillator. The PWM drives the active switch with a series of pulses at the regulated duty-cycle. [2]

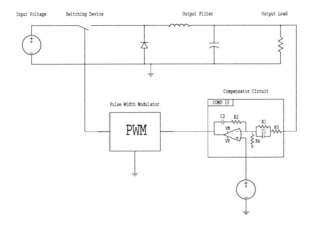

Figure 6.3 Voltage-mode buck converter schematic.

The forward converter is derived from the buck converter. It has the same basic operation as the buck converter; however, the forward converter provides isolation by means of a high-frequency transformer. See Figure 6.4 for a schematic of a voltage-mode double switch forward converter. The transformer reduces component stresses as a result of high frequency switching.[3] The double switch eliminates the need for a separate demagnetizing winding and also reduces component stresses.[4]

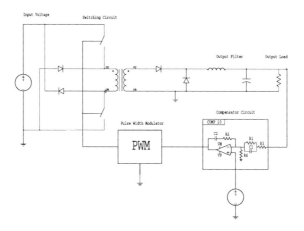

Figure 6.4 Voltage-mode double switch forward converter.

6.1.2 Model Implementation and Verification

Modeling a single switching power supply can be difficult if using a SPICE-like macro-model due to the limitations of abstraction capability. A device-level model will not be efficient with respect to simulation time. However, through an ABM the model implementation is made easier because there are no abstraction limits (as long as the behavior can be described mathematically).

Buck Converter

The easiest method to simulate the cascaded system is to analyze each power supply individually. First, we will analyze the buck converter. Once the buck converter has been designed (design of the output filter and compensation) it is good practice to verify that the control loop is stable. Basic design criteria for stability include:

1) no poles in the right-half side of the s-plane

2) there is adequate gain over the specified bandwidth for the open-loop characteristic

3) there must be at least 45 degrees of phase margin at the 0 dB open-loop crossover frequency

4) the steady-state error for the closed-loop system should be zero (i.e. no oscillations exist in steady-state) [5]

To analyze stability, averaged (state-spaced) models can be used. These models can be found in the Analogy model library.

See Figure 6.5 for a schematic of the buck converter model. In the schematic the averaged buck model takes the place of the PWM, the active switch, and the passive switch (diode). The model is based on the average current and voltage at the switched node (node between the output of the active switch and the inductor). The average voltage is calculated based on the duty-cycle of the converter. This specific model is written as a teaching tool only for buck voltage-mode converters designed to operate in the continuous conduction mode.[6]

The output filter is modeled by discrete component models; the compensation circuit is modeled with a basic operational amplifier model and discrete components. A two-pole/two-zero compensation scheme is chosen for this design. The *breakpoint* model allows the user to perform three types of analyses with the same schematic: open-loop turn-on transient (based on a specified duty-cycle), closed-loop turn-on transient analysis, and an open-loop small signal AC analysis. Because of the linear nature (i.e. no switching is involved) of the averaged model, an AC analysis can be performed on the control loop.[6] See Appendix A.1 for the netlist of the average buck converter model. All the values for model parameters can be found in the netlist.

Distributed Power Supply

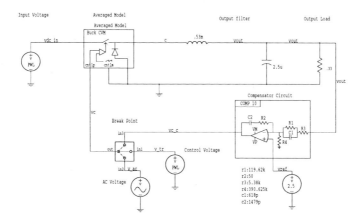

Figure 6.5 Buck converter model schematic.

A turn-on transient and small signal AC analysis were performed on the average buck model. The turn-on transient analysis is used to verify that the system is damped (i.e. over-damped, critically damped, under-damped, or oscillatory) for best system performance and that the converter is regulated properly. The AC analysis will help determine the degree of stability (i.e. phase margin, cross-over frequency, DC gain, etc.)

A 10 millisecond transient analysis was performed on the average buck model:

```
tr (siglist / /*.*/*,tend 10m,trip zero,tstep 1u).
```

The transient initial point, *trip*, was set to zero - this sets the initial DC conditions for all nodes in the circuit to zero.

To break the loop for the small signal AC analysis the *breakpoint* model parameter *input* had to be altered to *use2* (sets the switch to the AC input voltage source):

```
alter /switch_vin.breakpt1 = input=use2.
```

143

An *alter* allows the user to change the netlist without re-netlisting the schematic while using the Saber design tool.

The small signal AC analysis was performed from 1 hertz to 10 megahertz:

```
ac (acip tr,fbegin.1,fend 100meg).
```

The ac initial point, *acip*, was set to trigger off the end point of the transient analysis. At the end of the transient analysis the system is in steady-state, thereby allowing the open-loop to be perturbed about steady-state operation.

The results of both the transient and AC analyses can be found in Figure 6.6. The transient result, shown in the lower graph, reveals an overshoot of approximately 2 volts and a regulation voltage of 3.3 volts. The open-loop (V*c*/V*error*) gain/phase analysis in the upper graph reveals a phase margin of approximately 27 degrees.

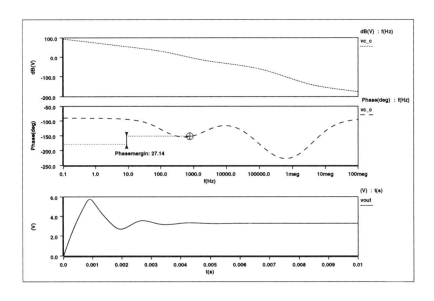

Figure 6.6 Averaged buck converter transient and AC simulation results.

Distributed Power Supply

The present system could potentially become unstable. To correct this problem the compensation circuit values must be changed to boost the phase around the zero crossover frequency. See Appendix A.1 for the changed compensation component values in the average buck converter netlist.

See Figure 6.7 for the simulation results of the improved system. The turn-on transient response reveals a reduced overshoot voltage and is also less oscillatory. Thus, the system has become more damped. The open-loop gain/phase graph reveals a phase margin of 83 degrees which is acceptable for basic stability requirements.

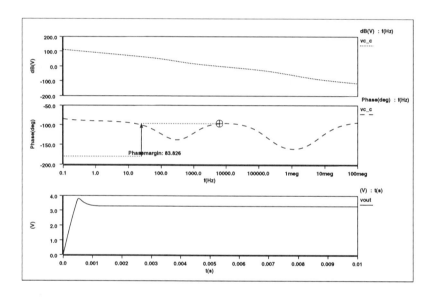

Figure 6.7 Improved averaged buck converter transient and AC simulation results.

The averaged buck converter model has allowed us to improve the loop stability of the converter through simulation. Performing such analysis in the laboratory is difficult and requires specialized equipment. Furthermore, implementation of an averaged macro-model for a buck converter is possible; however, such an implementation is difficult and requires the use of many controlled sources which add unnecessary rows and columns into the system matrix; thus making simulation less efficient with respect to simulation time.

Forward Converter

Implementation of the forward converter model is very similar to the buck model. Again, we will examine the *averaged* behavior of the forward converter. See Figure 6.8 for the forward converter model schematic. The averaged model contains detail of the active and passive switches and the transformer gain of the forward converter architecture. The discrete components and compensation models are implemented just as they are in the buck converter model. Again, a two-pole/two-zero compensation scheme was chosen. The *breakpoint* model is also used so that the control loop can be broken to perform a small-signal AC analysis using a single schematic.

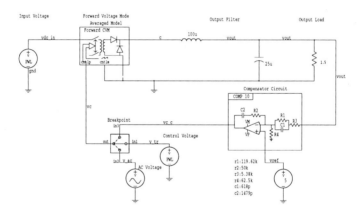

Figure 6.8 Forward converter model schematic.

A transient and AC analysis were performed on the model. The simulation results of the forward converter model can be found in Figure 6.9. The transient result reveals an overshoot of 2 volts and a regulation voltage of 15 volts. The open-loop AC analysis reveals a phase margin of approximately 8 degrees. Again, the chosen values for this compensation scheme could pose

Distributed Power Supply

stability problems (e.g. oscillations). To demonstrate how this could be a problem with the overall system, these compensation values will be used in the simulation of the cascaded system.

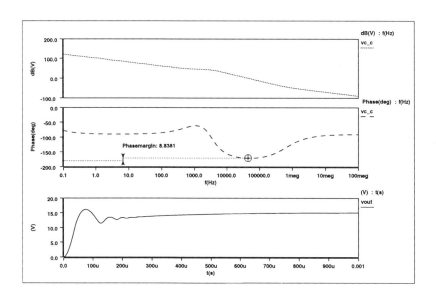

Figure 6.9 Averaged forward converter transient and AC simulation results.

Cascaded Model

Once the individual converters have been verified, they can then be cascaded to verify a link of the distributed system. See Figure 6.10 for a model schematic of a cascaded system. In this model, the rectified DC input is represented by a DC voltage source. The output of the forward converter directly feeds the input of the buck converter. The load is represented by a 3.3 ohm resistor that will produce 10 amps. See Appendix A.3 for the cascaded converter netlist.

As can be observed from the netlist, model hierarchy was established in the cascaded model. The hierarchy is established by creating a sub-circuit model. This makes the top-level schematic cleaner and helps to keep the design more organized. The sub-circuit is created by identifying the input/output pins on the sub-circuit schematic. The netlister recognizes the hierarchy from the top-level schematic and partitions the netlist into intermediate templates and a top-level netlist. Remember that a netlist can be established within a model template when using MAST.

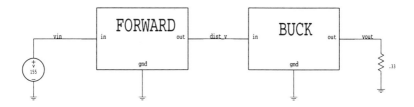

Figure 6.10 Cascaded converter model schematic.

To further demonstrate the ability to model systems at different abstraction levels, each converter for this example will be modeled in more detail.

Expanded Buck Converter Model

The buck converter model will be expanded to include the active and passive switches. Hence, the switching behavior of the converter will be introduced into the total cascaded model. The active switch will be modeled as an ideal switch and the diodes will be modeled as ideal diodes. The PWM will be modeled by actual logic and an oscillator that will produce the clock and the sawtooth waveforms. The compensation model and output filter components have not been changed in this model.

Another interesting addition to this model is the *duty-cycle* measurement template. AHDL's make it easy to write models that can take various measurements (i.e. rise-time, duty-cycle, power dissipation, etc.) within a system. The *duty-cycle* measurement template monitors any digital node and computes the duty-cycle (0-1) based on the ratio: high-time to the period of the input. See

Distributed Power Supply

Figure 6.11 for the expanded buck converter sub-circuit model schematic. See Appendix A.3 for the values of the model parameters in the buck converter sub-circuit model.

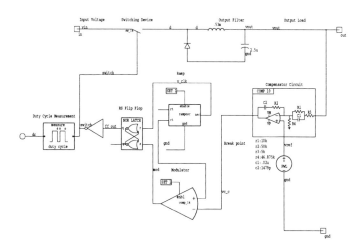

Figure 6.11 Expanded buck converter sub-circuit model schematic.

Expanded Forward Converter Model

The forward converter model has been expanded in much the same method by which the buck converter model was expanded - active and passive switches, PWM logic and oscillator, and transformer models have been added. The compensation model and output filter components have not been changed. See Figure 6.12 for the expanded forward converter sub-circuit model schematic. See Appendix A.3 for the values of the model parameters in the buck converter sub-circuit model.

149

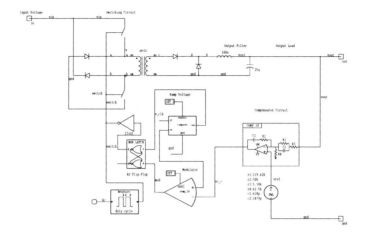

Figure 6.12 Expanded forward converter sub-circuit model schematic.

Cascaded Converter Model Simulation and Results

Having completed the detailed models for the individual converters, the entire cascaded system can now be simulated. Remember that we are using a potentially unstable forward converter stage in this model.

A DC and 3 millisecond transient analysis were performed on this model:

```
dc
den 128
ter 1u
tr (siglist / /*:*/*,tend 3m,trip zero,tstep 1u).
```

Simulating switching power supplies can be very tricky and sometimes users can encounter inaccurate results if not careful. To help the simulator, *den* has been set to 128 and *ter* has been set to 1u. Increasing *den* will increase the number of simulation points in the simulation and decreasing *ter* will decrease the truncation error that will be allowed for simulation convergence. There are other simulator variables that can be changed to alter the simulation

Distributed Power Supply

accuracy - for more details on these simulation parameters please see the Saber Simulation Reference Manual.

The cascaded model simulation results can be found in Figure 6.13. The upper graph shows the output of the forward converter, *dist_v*; the middle graph shows the output of the buck converter, *vout*. The forward converter voltage regulates to 15 volts with approximately a 2 volt peak-to-peak oscillation. The buck converter output voltage regulates to 3.3 volts. The buck converter demonstrates good line-regulation, as the oscillation problem of the forward converter does not affect the operation of the buck converter. However, this could still pose other problems with the other stages in the distributed system.

The lower graph shows the output of the *duty-cycle* measurement template in the buck converter sub-circuit model. This result reveals that the maximum allowed duty-cycle (set by the PWM) is adequate for the buck converter to reach the specified regulation voltage.

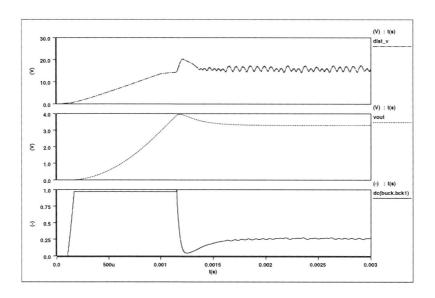

Figure 6.13 Cascaded converter transient simulation results.

Since the unstable forward converter stage could pose problems, we will correct it by altering the compensation values in the forward converter subcircuit. Revisiting the forward converter averaged model, we will change compensation values and re-run the transient and AC simulations. See Appendix A.2 for the new compensation values. See Figure 6.14 for the simulation results of the improved forward converter. The turn-on transient response reveals a result that is less oscillatory. Thus, the system has become more damped. The open-loop gain/phase graph reveals a phase margin of 59 degrees.

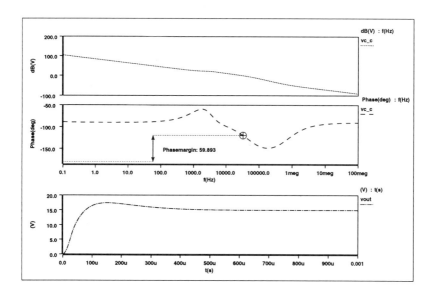

Figure 6.14 Improved averaged forward converter transient and AC simulation results.

With the improved forward converter control loop, we will again simulate the cascaded system. See Figure 6.15 for the transient simulation results of the improved system. The upper graph reveals a forward converter output voltage of 15 volts (without oscillations); the lower graph reveals the buck converter output voltage of 3.3 volts. The lower graph shows the buck converter duty-cycle which is cleaner due to the improved stability of the forward converter.

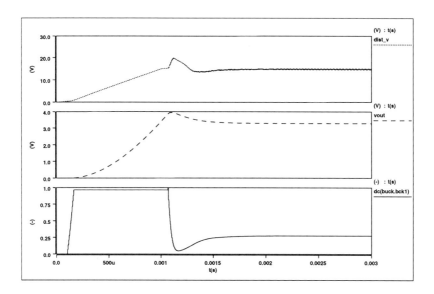

Figure 6.15 Improved cascaded converter transient simulation results.

Through modeling and simulation of a power supply system we have been able to obtain valuable design information.

These models can be used to employ the hierarchical design methodology. The averaged models are system abstractions that can be implemented to analyze architectural trade-offs, benchmark performance, and optimize system performance. When designing top-down, more detail can be added to expand the models as was done with the switching version of the converter models. At the lowest-levels of abstraction, the operational amplifier in the compensation circuit could be modeled at the device-level; the device-level design could be optimized for the best control loop-stability. Finally, design information gained at the device-level can be used to characterize the higher-level models; hence bottom-up verification is achieved.

153

6.2 Automotive Ignition System

In this example, an automotive ignition system is modeled and simulated. It demonstrates implementation of a basic behavioral macromodel for a mixed technology IC system. Specifically, a sparkplug ignition will be analyzed using different ignition coil charging (dwell) times.

6.2.1 System Overview

A basic electronic ignition system contains the following blocks: an ignition switch, solenoid/relay, starter motor, ignition coil, ignition coil driver, distributor, crankshaft position sensor, spark plugs, and a battery. The purpose of an automotive ignition system is to ignite a compressed air/fuel mixture in the motor cylinders. See Figure 6.16 for a simplified block diagram of an electronic ignition system; the arrows indicate the signal path.

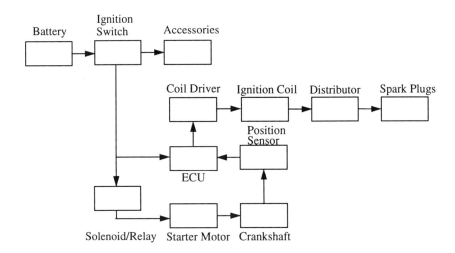

Figure 6.16 Block diagram of electronic ignition system.

System operation is commenced when the ignition switch is turned on and the accessories are activated. As the switch turns to the start position, the solenoid/relay is activated; this completes the circuit between the starter motor and the battery. As the engine turns, the position sensor (magnetic reluctor signal generator), reports information of crankshaft position to the electronic control unit (ECU). The ECU will then send a pulse signal to the coil driver circuit which will enable the ignition coil (transformer). As the current flows through the primary winding, a magnetic field is created in the core of the coil. Energy is then transferred from the primary to the secondary coil until the voltage is great enough (approximately 15 kV to 20 kV) to produce an arc across the spark plug electrodes. This charging time is known as the dwell period. The voltage in the secondary winding must be delivered to each spark plug at a specific time; the distributor is the ignition component which serves this purpose. [7]

6.2.2 Model Implementation and Verification

The model implementation is not too difficult. In this example, all the main blocks of an electronic ignition system are implemented in the model for the exception of the distributor block. For simplicity, only one spark plug is modeled. See Figure 6.17 for the top-level electronic ignition system model schematic.

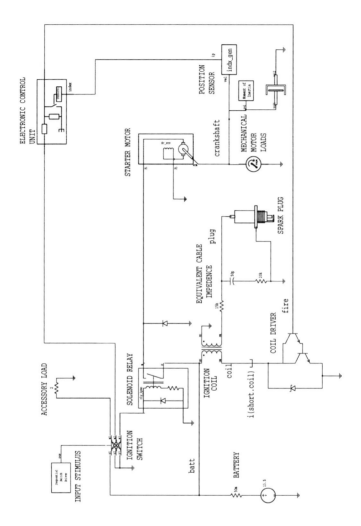

Figure 6.17 Top-level automotive ignition system model schematic.

The following are short descriptions of all the subsystem models contained within the top-level ignition system model:

Ignition Switch

The ignition switch model, *sw_mtrx3* (multi-position matrix switch), is taken from the Analogy model library. This model has three inputs, three outputs, and a control input that specifies which set of matrix switch positions (i.e. which input is connected to what output) to use. The matrix switch positions are specified by the user defined parameter, *mtrx*. The parameter settings can be found in Appendix A.4.

Solenoid/Relay

The solenoid/relay model is also taken from the Analogy model library. The model *rly_1pno* (single-pole, normally-open relay) is used. This model has differential inputs and outputs. The model contains: an input diode with a parallel inductance coil and an equivalent series resistance and an output switch that is activated by the input voltage. The *on* and *off* resistances of the switch, diode device parameters, inductance and equivalent resistance are all specified by user defined parameters. The parameter settings can be found in Appendix A.4.

Ignition Coil

The ignition coil is modeled by a 2-winding transformer, *xfr*, with a magnetic core. The model allows the user to specify various transformer parameters (e.g. primary and secondary inductance, turns ratio, cross sectional core area, e.t.c). The parameter settings can be found in Appendix A.4.

Battery

The battery is simply modeled by a 13.5 volt ideal voltage source with a 50 milli-ohm series resistor. A more complex model could be implemented. However, in this example, a simple model is adequate to demonstrate the basic functionality of an automotive ignition system.

Starter Motor

The starter motor model can also be found in the Analogy model library. The DC motor model *dc_srs* is used. This model is a permanent magnet motor model with parasitic electrical resistance and inductance. The model has two electrical inputs and one mechanical output. See Appendix A.4 for the user defined parameter settings.

Electronic Control Unit

The ECU is modeled by a switch that generates a pulse that is initiated by the position index input (output of the position sensor). The length of the pulse is determined by the user defined parameter, *dwell*. The dwell time is the amount of time the ignition coil will charge. The pulse signal generated by the ECU drives the base of a Darlington configuration that in turn drives the ignition coil. See Appendix A.4 for the user defined parameter settings of the ECU and Appendix A.5 for the ECU model code.

Motor Crankshaft

The motor crankshaft is simply a lumped model of mechanical loads. The mechanical loads include: viscous friction, and inertial loads. A torque source is used to model the torque that is applied to the crankshaft, by the engine, after the engine has been started. See Appendix A.4 for parameter values of the mechanical loads and torque source.

Position Sensor

The position sensor models the magnetic reluctor signal generator. The template name is *indx_gen* and has two pins: an angular velocity input pin and an index output pin. This model derives the spark-fire timing which is based on the crankshaft angle. The angular displacement of the crankshaft is calculated by differentiating the angular velocity of the crankshaft. If the crankshaft angle is greater than the specified threshold (.5 in this example), the output index pulse is set to 1, otherwise, it is 0. See Appendix A.6 for the position sensor model code. Note this model contains an angle initializing parameter, *thetainit*, that is set internally in the model template to 0.

Coil Driver

The coil driver is modeled by a pair of *q2n6043* bipolar transistor models. These transistors are implemented in a Darlington configuration. A *d1n3012* zener model is placed in parallel with the Darlington configuration. See Appendix A.4 for the placement of the coil driver with respect to the entire system.

Spark Plug

The spark plug model is a customized model, *sparkplug*. This model contains 3 user defined parameters: breakdown voltage *vbrk*, minimum arc sustaining voltage *vsus*, and the resistance of the plug during spark *rarc*. The sparkplug model has one input and one output. Once the input voltage increases above the specified arc voltage, the sparkplug will break down and the terminal impedance will change from the nominal no-arc resistance to the arc resistance, thereby, discharging the sparkplug. See Appendix A.7 for the sparkplug model code and Appendix A.4 for the user defined parameter settings.

Simulation and Results

A DC and a 4.2 second transient analysis were performed on the ignition model:

```
dc
sigl batt crankshaft angvel_fps(damper_w.damper)\\
     plug i(short.coil) fire
di
te 5
ts 1m
ter 100u
tsmax 20m
pf dwell_2m.
```

The above commands were set up in a run file that can be executed in Saber. Various simulation parameters (*te*, *ts*, *ter*, and *tes*) were set in this file along with the specification for the type of simulation analysis. The command *pf dwell_2m* specifies the name output file where the simulation results will be contained.

The stimulus for the test circuit is a user-programmed stream of switch position values that is user defined in the *sdr_prsq* switch driver model. The following table shows what switch position is invoked at various time steps:

time (seconds)	switch position
0	all switches are *open*
1	switch *S11* is *closed* (accessories on)
2	*S11* and *S12* are *closed* (accessories and ECU on)

time (seconds)	switch position
2.2	*S11 open*, *S12* and *S13 closed* (accessories off, ECU and ignition relay on)
4	*S11* and *S12 closed* (accessories and ECU on)

See Appendix A.4 for the input stimulus connectivity, and the user defined parameter settings for the *sdr_prsq* model.

See Figures 6.18 and 6.19 for the electronic ignition transient simulation results. In Figure 6.18 the upper graph reveals the sparkplug voltage; the middle graph reveals the battery voltage; the lower graph shows the crankshaft position. Initially, as the accessories are turned on, the battery voltage decreases. The accessories are turned off before the starter motor is engaged through the solenoid/relay. As the engine starts, the battery voltage begins to decrease until the first sparkplug firing. Note that as the crankshaft increases in angular position, the sparkplug firing becomes more frequent and the battery voltage eventually recovers as the engine takes over the rotation of the crankshaft.

Automotive Ignition System

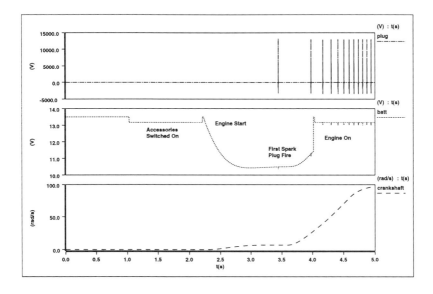

Figure 6.18 Electronic ignition transient simulation results: spark plug voltage, battery voltage, and crankshaft position.

In Figure 6.19, a zoomed-in graph of the first sparkplug firing is revealed: the upper graph shows the *fire* voltage which is the base voltage of the Darlington pair; the middle graph reveals the ignition coil primary current; the lower graph reveals the sparkplug voltage. The first event to note is the dwell period (2 milliseconds). The dwell period begins as the *fire* voltage causes the ignition coil to charge to approximately 2.3 amps. Once the *fire* voltage is released, energy is transferred to the secondary of the ignition coil igniting the sparkplug at the 13 kilo-volt break down. The duration of the spark is about 1 millisecond. As the spark is extinguished, the residual energy in the ignition coil is dissipated.[8]

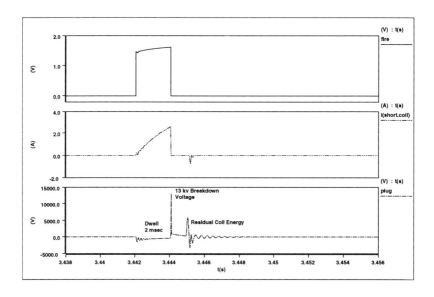

Figure 6.19 Electronic ignition transient simulation results: *fire* **voltage (base of Darlington pair), primary side of ignition coil current, and sparkplug voltage.**

To demonstrate the effect of changing the dwell time on the sparkplug ignition, a second DC and transient simulations were performed on the electronic ignition model with a modified dwell time of 0.5 milliseconds:

```
alter /ecu_ign.eng = dwell = 0.5m
dc
sigl batt crankshaft angvel_fps(damper_w.damper)\\
     plug i(short.coil) fire
di
te 5
ts 1m
ter 100u
tsmax 20m
pf dwell_05m.
```

The results of the modified dwell time simulation can be found in Figure 6.20. The upper graph reveals the sparkplug voltage with the 0.5 millisecond dwell time; the lower graph is the sparkplug voltage with a 2 millisecond dwell time. Note that the sparkplug voltage with the decreased dwell time does not ignite, because the secondary peak voltage does not exceed the breakdown voltage; energy in the coil is immediately dissipated after the *fire* voltage is released.

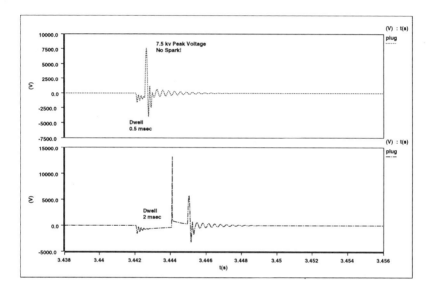

Figure 6.20 Electronic ignition transient simulation results with modified dwell time.

In this example, a variety models have been used in a top-level analysis of an electronic ignition system that contains a variety of mixed-technology blocks. Further analyses can be performed on this system that can assist in the optimization of the ignition system before beginning detailed design. Again, the modeling continuum can be traversed (top-down) by modeling the individual components in more detail. Once the lowest-level of required detail is achieved, the higher-level models can be characterized to incorporate second-order effects; hence the designed can be verified bottom-up.

6.3 Audio Test System

In this example, an audio test system is modeled and simulated. This example demonstrates how to implement a basic behavioral macromodel of another mixed technology IC system. Specifically, the performance of an audio system is analyzed with and without a notch-filter that is used to suppress the mechanical resonance of a loudspeaker.

6.3.1 System Overview

A basic audio test system consists of: a test tone generator, a digital signal processor (DSP), a crossover network to separate out the various frequency ranges for each speaker (i.e. tweeter, mid-range, and bass), a sensing mechanism (microphone), and a frequency analyzer that can determine frequency and phase response. The test tone generator is used to generate various waveforms (e.g. impulse, tone burst, e.t.c) to test the audio system. The DSP is used to create special sound effects and to compensate the mechanical speaker system for better performance. See Figure 6.21 for a simplified block diagram of an audio test system. [9] The arrows indicate the signal path.

Audio Test System

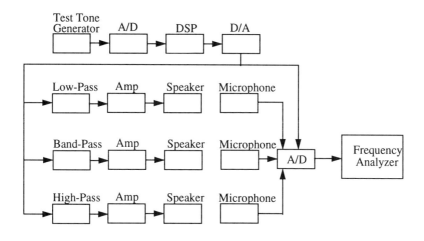

Figure 6.21 Audio Test System Block Diagram.

The test tone generator creates the test patterns that excite the audio system. The stimulus is then converted to digital form so that special effects and compensation can be performed digitally. The output of the DSP is then converted back into an analog signal where it is filtered by a crossover network. The low, medium, and high frequencies are then applied to the appropriate loudspeaker via an amplifier capable of handling specified frequencies. A microphone is used to sample the audio output of the loudspeakers. The sampled data is then converted to a digital signal for processing. From this data, frequency response information can be extrapolated via an Fast Fourier Transform (FFT) analysis. The performance of the audio system can then be analyzed for distortion, noise, and clarity.

6.3.2 Model Implementation and Verification

For purposes of demonstration, only the low-tone loudspeaker and supporting circuitry is modeled. The top-level model blocks include: test tone generation block, clock generation block, analog-to-digital conversion block, digital signal processing block, crossover filter, power amplifier block, and loudspeaker block. See Figure 6.22 for the top-level model schematic of the audio test system and Appendix A.8 for the top-level netlist. This netlist contains the connectivity and settings for all the user defined parameters for both the top-level and all respective sub-systems.

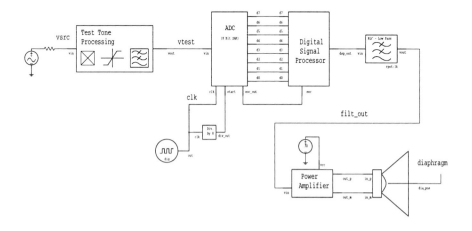

Figure 6.22 Top-level audio test system model schematic.

The top-level block will function as follows: the test patterns are generated by the test tone generator; the 8-bit ADC converts the data to digital bits; the DSP has two functions: first, it is used to implement a notch-filter that will suppress mechanical resonances of the loudspeaker; and second, it is used to implement an echo sound effect; a crossover (low-pass filter) is then used to filter the high frequency harmonics in the signal; and finally, the signal is applied to a power amplifier which drivers the loudspeaker.

The functions used to sense the loudspeaker output (microphone) and process the data do not have to be modeled, because the output can be post-processed. This means an FFT analysis, or any analytical analysis, can be applied directly to any signal via the post-processor of the simulator. This information can then be used to study of performance and optimization of the system.

Audio Test System

The following are individual descriptions of the various audio test system blocks:

Test Tone Generator

The input to the test tone generation block is a sinusoidal voltage source. This signal is then converted into a control signal by an *electrical-to-control* model, where it is multiplied by a pulse wave to produce a tone burst of 50 milliseconds. The signal is then limited to a plus and minus 5 volts via a limiter model. At the output of the test tone generator, the signal is smoothed by a 100 Hz single-pole lag filter. The output stage converts the control signal back into an electrical signal through a *control-to-electrical* converter model. All the blocks in this model are part of the standard Analogy model library. See Figure 6.23 for a schematic of the test tone generator model.

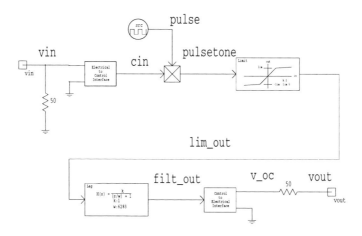

Figure 6.23 Test tone generator model schematic.

Clock Generator

The clock generator sub-system provides the *clk* signal in top-level model. The clock generation sub-system consists of a ring oscillator that is made up of an NAND gate and buffers with long delay times. The delays were derived from actual analog circuits that have been characterized at the device-level.

167

The ring oscillator is initiated by a programmable bit stream source that is set to produce a logic level 0 at 0 seconds and transitions to logic level 1 at 1 picosecond. This model is very efficient with respect to simulation time and contains the necessary amount of abstraction to demonstrate the basic functionality of an audio test system. See Figure 6.24 for a model schematic of the clock generator block.

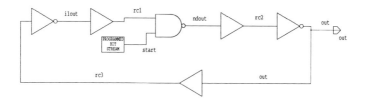

Figure 6.24 Clock generator model schematic.

Analog-to-Digital Converter (ADC).

The ADC in this system is implemented for demonstration only; the analog data could be directly converted to the state domain through a analog to z-domain converter model. The 8 bit ADC consists of: a successive approximation register (SAR), a digital-to-analog converter (DAC), a comparator, a shift register, and a buffer. The shift register, comparator, and buffer are analog behavioral models from the Analogy model library. The DAC is a behavioral macromodel consisting of switches, resistors, and an operational amplifier. The SAR model code can be found in Appendix A.13. See Figure 6.25 for a model schematic of the ADC block.

Audio Test System

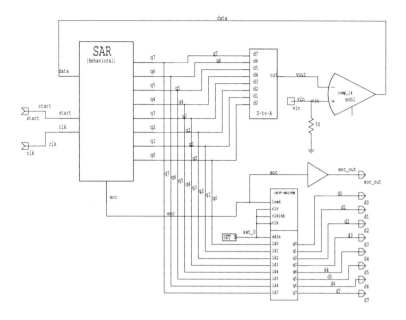

Figure 6.25 Analog-to-digital converter model schematic.

Digital Signal Processor (DSP)

The DSP front-end converts an 8-bit binary digital word into a z-domain signed integer via the *b2z* model. The reason for this conversion is that if digital words (bytes) are used, the actual logic needed to process the individual bits would have to be implemented in the model; this logic would include: memory, multiply/accumulate logic, firmware to drive the logic, etc. Instead, the DSP is abstracted by using z-domain blocks (similar to s-domain transfer function blocks). A z-domain transfer block is used to implement the notch-filter used to suppress the mechanical resonance. A pure delay block is used to create the echo effect; the output of the delay is added back into the main signal through a summation block. Gain blocks are also used to adjust the amplitudes of the original and delayed signals independently.

Once the data has been summed, it passes through a z-domain-to-analog converter model; through this converter the actual digital-to-analog conversion is abstracted. A digital to z-domain model, *clk2smp*, is used to produce the sample clock for all the z-domain elements from the external digital signal *eoc*. See Figure 6.26 for the DSP model schematic and Appendix A.14 for the clock-to-z-domain converter model, *clk2smp*.

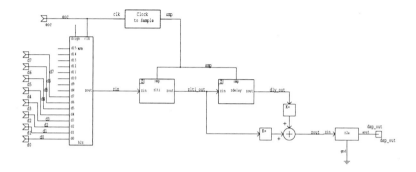

Figure 6.26 Digital signal processor model schematic.

Crossover Filter

The crossover filter is a basic RLC low-pass 2nd-order filter. All the models were taken from the Analogy model library. See Figure 6.27 for a model schematic of the crossover filter.

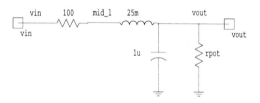

Figure 6.27 Crossover low-pass RLC filter.

Power Amplifier

The power amplifier block consists of a single stage common-emitter amplifier with output transformer for isolation. All the models were taken from the Analogy model library. See Figure 6.28 for a model schematic of the power amplifier.

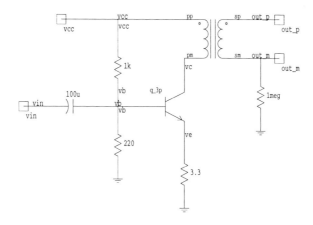

Figure 6.28 Power amplifier model schematic.

Loudspeaker

The loudspeaker model consists of an electrical-to-mechanical transducer block and lumped mechanical model. The transducer model, *voice_coil*, converts electrical energy to mechanical translational energy. See Appendix A.11 for the *voice_coil* model. The lumped mechanical components model the diaphragm. The diaphragm model is very basic. Depending on the diaphragm enclosure, shape and stiffness in this model can vary. The components of the lumped model include: spring, mass, and wind-drag. The diaphragm stiffness is modeled by the spring; the diaphragm inertia is modeled by the mechanical mass, and the diaphragm load is modeled by wind-drag. The spring model, *spring_nl*, can be found in Appendix A.15. The spring model is a non-linear model that represents a non-linear suspension system; at different displacements the stiffness can vary. The wind-drag model, *winddrag*, can be found in

Appendix A.12. The winddrag model produces a non-linear damping force that is proportional to the translational velocity; the non-linearity is a result of a changing winddrag force at various diaphragm velocities. [10] [11] See Figure 6.29 for the loudspeaker model schematic

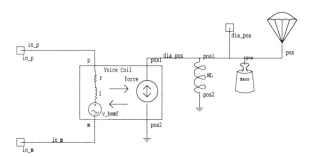

Figure 6.29 Loudspeaker model schematic.

Simulation and Results

Before simulating the entire system, it is sometimes advantageous to first work on the subsystems. We will first examine the frequency response of the loudspeaker. See Figure 6.30 for the loudspeaker subsystem test schematic and Appendix A.9 for the loudspeaker subsystem netlist.

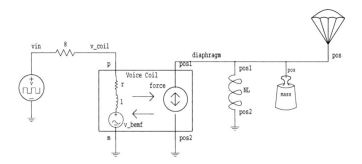

Figure 6.30 Loudspeaker subsystem test schematic.

Audio Test System

A DC and AC small signal analyses were performed on the loudspeaker subsystem; the input voltage source will be the trigger for the small signal analysis:

```
dc
ac (pf ac, fb 1, fe 1k, np 1024)
```

The results are shown in Figure 6.31. The results reveal a mechanical resonance peak at 57 hertz. It should be noted that the AC analysis does not consider non-linearities due to the damping effects of air resistance and diaphragm stiffness, but only the initial values are considered.

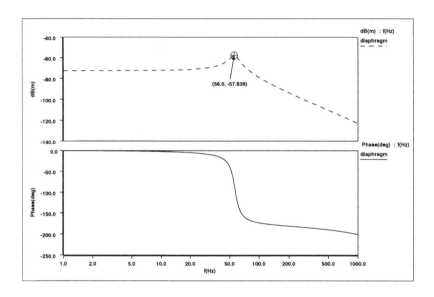

Figure 6.31 Loudspeaker subsystem small signal AC results.

Based on the determined mechanical resonance frequency, the notch-filter can be designed using the DSP sub-system. Before analyzing the entire audio system, the DSP subsystem can examined to ensure that the proper frequencies are suppressed. Figure 6.32 shows the DSP subsystem test schematic and

Appendix A.10 shows the DSP sub-system netlist.

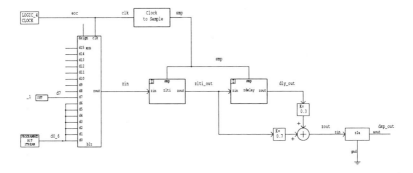

Figure 6.32 DSP subsystem test schematic.

This test includes the ADC with a digital input set to produce an impulse with a magnitude of 130 and a width of 400 microseconds. From this stimulus, the impulse response of the *zlti* z-domain notch-filter block can be obtained. The FFT of the impulse response can then be processed to obtain the frequency response of the digital filter.

A DC and transient analyses were performed on the DSP subsystem:

```
dc
tr (pf tr, te 200m, ts 50u, mon 300).
```

The FFT was invoked by the following command:

```
fft (pfin tr, pfout fft, cn zlti_out, axis log).
```

The impulse response results of the z-domain filter in the DSP can be found in Figure 6.33. The upper graph shows the input to filter block and the lower graph is the impulse response.

Audio Test System

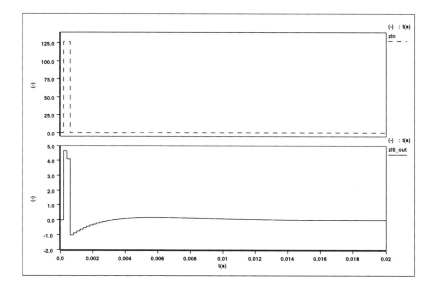

Figure 6.33 DSP subsystem results: impulse response of the z-domain filter.

The FFT results of the impulse response can be found in Figure 6.34. The notch characteristic is revealed at approximately 60 hertz. The notch will help suppress mechanical loudspeaker resonance at 56 hertz. This type of analysis can be of value to the designer when attempting to integrate the entire system to achieve the best system performance.

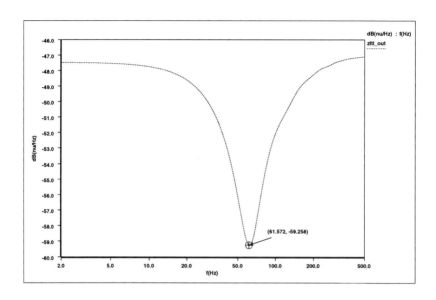

Figure 6.34 DSP subsystem results: FFT of digital filter impulse response.

Now that we have verified the DSP subsystem, we can confidently simulate the entire audio test system. For the following top-level test-setup explanation refer to Figure 6.22.

The input stimulus is produced by a 100 Hz 7 volt peak-to-peak input voltage source. The test tone generator produces a 100 hertz 50 millisecond tone burst. This tone burst is then converted into a 8 bit digital word via the ADC. The digital word is then converted into a z-domain event signal that is filtered and echoed in the DSP. The signal is then applied to the low-pass crossover filter where the high frequency harmonics are attenuated. Finally, the signal is amplified and converted into airwaves via the transducing loudspeaker.

The first system test consists of DC and transient analyses performed with the DSP filter is disabled:

Audio Test System

```
#disable filter with alter command
a dsp.dsp1/zlti.zlti1/a=0.039
a dsp.dsp1/zlti.zlti1/num=[1,0,0]
a dsp.dsp1/zlti.zlti1/den=[1,0,0]
dc
tr (pf tr_nofilt, df _, te 200m, ts 1u, mon 1000,\\
    ter 100u
```

The transient results for the top-level audio system with disabled filter can be found in Figure 6.35. The upper graph is the input, *vrsc*. The second graph from the top is the tone burst signal, *vtest*. The third graph is the low-pass filter output, *filt_out*. The lower graph is the loudspeaker output signal, *diaphragm*. The *diaphragm* signal represents the instantaneous position of the loudspeaker diaphragm measured in meters. Note that the loudspeaker output is not very clean. However, the echo effect is present in the *diaphragm* signal.

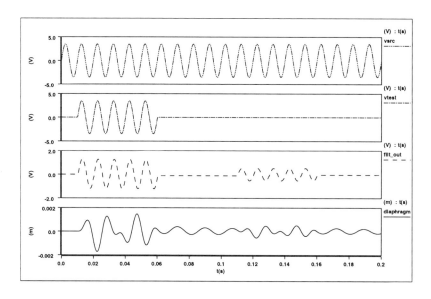

Figure 6.35 Top-level audio system: transient results with disabled filter.

Now we will run DC and transient simulations where the DSP filter is enabled:

```
#enable filter with alter
a dsp.dsp1/zlti.zlti1/a=0.036706
a dsp.dsp1/zlti.zlti1/num=[1,-1.981254,0.986114]
a dsp.dsp1/zlti.zlti1/den=[1,-1.864734,0.869309]
dc
tr (pf tr_filt, df _, te 200m, ts 1u, mon 1000, ter 100u
```

The transient results for the top-level audio system with enabled filter can be found in Figure 6.36. These results are compared against the results of the previous simulation. The upper two graphs are *filt_out* and *diaphragm*, respectively, of the top-level system with the notch-filter enabled. The lower two graphs are *filt_out* and *diaphragm*, respectively, of the top-level system with the notch-filter disabled. From these results it can be inferred that the notch-filter helps to reduce the amount of noise and distortion in the loudspeaker output signal.

Audio Test System

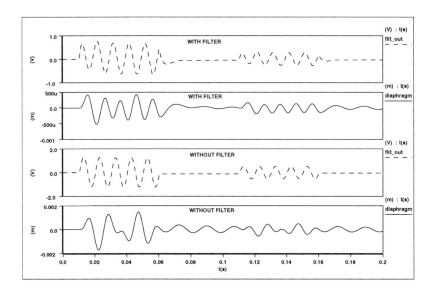

Figure 6.36 Top-level audio system: comparison of transient results with filter enabled and disabled.

For further analysis, we can take the FFT of the signal *filt_out* for both transient simulations. The FFT results can be found in Figure 6.37. In both waveforms we can observe the 100 hertz tone burst. However, in the signal that was obtained from the simulation with the notch-filter enabled, harmonics around 55 hertz are clearly suppressed.

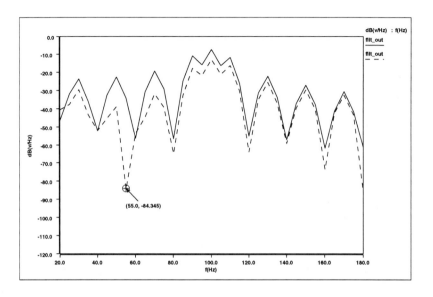

Figure 6.37 Top-level audio system: FFT results of with and without filter.

In this example, it is shown how a mixed-level, mixed-technology, and mixed-signal system can be integrated through modeling and simulation such that overall system performance is optimized. We have demonstrated various analyses at both the top and subsystem levels of a complex system. Again, we have approached design from the top. After all the top-level issues have been investigated, the individual blocks can be designed at lower-levels of abstraction. For example, once all the DSP requirements have been understood in the audio test system, that block can be designed top-down to the device-level. After an understanding of how the DSP will function at the device-level, the lower-level second-order effects can be incorporated in the higher-level models; thus the system can be verified bottom-up.

6.4 Digital Communication System

In this example, a digital communication system will be modeled and simulated. It will demonstrate how basic behavioral blocks can be used to model an broadband modulation/demodulation system. Specifically, a quadrature phase-shift keying (QPSK) modulation/demodulation system will be modeled and simulated. The performance of the system will be analyzed using two different settings of the signal pole low-pass filter in the demodulator stage.

6.4.1 System Overview

QPSK transmission is the concept of transmitting symbols via two carriers, in phase and quadrature (I-Q), to one another (i.e. $\sin wt$ and $\cos wt$). A symbol is a set of two successive binary pulses that can result in four constellation pairs: -1-1, -11,1-1, and 11 which correspond to the logic values 00, 01, 10, and 11 respectively. Hence, the modulated signal can consist of four different phases: 0,+ 90, -90, and 180 degrees.

Since each symbol represents a group of two bits, the symbol rate is one-half that of binary phase-shift keying (BPSK) modulation/demodulation; therefore, the spectrum frequency spectrum needed for QPSK transmission is one-half that of BPSK. QPSK has applications in modems, satellite communications, global positioning systems, and is also used in code division multiple access (CDMA) telecommunication systems. [12]

A basic QPSK communication system consists of: a serial-to-parallel (S-to-P) data converter, low-pass (LPF) band-limiting input modulator filters, a quadrature modulator, a transmission medium, a quadrature demodulator, more low-pass filters which provide harmonic filtering and bring the signal back down to base-band, a data slicer that further shapes the signal back into binary data, and a parallel-to-serial converter.

For the following explanation, see Figure 6.38 for a simplified block diagram of a QPSK modulation/demodulation system. The S-to-P block creates the symbol from the serial bit stream. The low-pass filters are optional and serve to band-limit the transmitted spectrum. The symbol components are then applied to quadrature modulator which is capable of carrying phase and amplitude information. Once the symbol components are modulated, they are summed together to form the signal which will be transmitted via the transmission medium.

On the demodulator side, the received signal is applied to a quadrature demodulator where the individual components of the symbol are demodulated by the same orthogonal carriers by which they were modulated. The demodulated signals are then fed through low-pass filters which filter higher harmonics and bring the I-Q signals back down to base-band. The filtered signals are then passed through a data slicer which recovers the original logic bit-stream via a comparator. Once the bit-stream has been recovered the symbol is then converted back into a serial bit-stream via a parallel-to-serial (S-to-P) converter. [12][13]

Digital Communication System

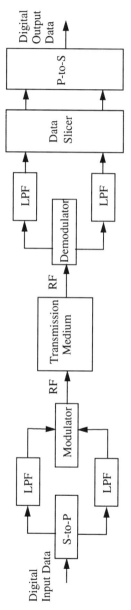

Figure 6.38 QPSK communications system.

183

6.4.2 Model Implementation and Verification

The model implementation of the QPSK modulator/demodulator is straightforward. For simplicity, the S-to-P and P-to-S converters will not be implemented in the model. The transmission medium and modulator low-pass filters were also omitted. Modeling the quadrature modulator/demodulator and data slicer blocks will contain enough abstraction to demonstrate the basic functionality of the QPSK communication system. See Figure 6.39 for a top-level model schematic of the QPSK modulator/demodulator example.

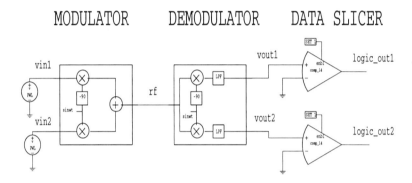

Figure 6.39 Top-level QPSK modulator/demodulator model schematic.

In this example, the modulator and demodulator are hierarchical. The data slicer was implemented using ideal comparators. A 0 volt compare voltage was applied at the minus input to the comparators. In the actual system, the compare voltage may vary according the DC levels in the demodulated signal.

The following are descriptions of the blocks contained within the QPSK digital communications system model:

QPSK Modulator

The QPSK modulator is created from models found in the Analogy model library. See Figure 6.40 for a model schematic of the modulator block. The carrier signals are produced by two 10 kilo-hertz 1 volt peak-to-peak voltage

Digital Communication System

sine sources. One of the sources is set with a 90 phase delay. The mixers are standard multiplier blocks; the summer is also a standard block. Note that the input signal is converted from a voltage to a control signal via the *electrical-to-control* block and then converted back into a voltage via a *control-to-voltage* converter block. See Appendix A.16 for the settings of the user defined parameters and the model netlist.

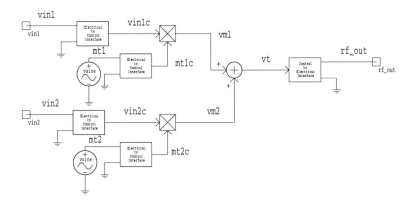

Figure 6.40 QPSK modulator model schematic.

QPSK Demodulator

The QPSK demodulator block is similar to the modulator block. See Figure 6.41 for a model schematic of the quadrature demodulator block. The demodulator block contains the low-pass filters that are implemented as transfer function blocks where the numerator and denominator are defined by the model parameters. See Appendix A.16 for the settings of the user defined parameters and the model netlist.

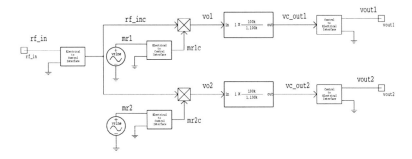

Figure 6.41 QPSK demodulator model schematic.

Simulation and Results

The test setup for the QPSK modulator/demodulator system can be found in Figure 6.39. The stimuli are two piecewise-linear voltage sources set to cycle through three symbol combinations: (-1,1),(1,1), and (-1,-1). The first bit is set on node *vin1* and the second bit on node *vin2* (1st bit, 2nd bit). Each symbol combination is set for 200 microseconds before cycling through the next symbol set. For simplicity, the demodulator low-pass filters are defined as a single-pole set at 100 kilohertz and with a DC gain of 100 dB.

A DC and a 1.2 millisecond transient analyses were performed on the test setup:

```
dc
tr (siglist / /*.*/*,tend 1.2m, tstep.01u).
```

In Figure 6.42 the internal modulator block transient results can be found. The lower two graphs show signal *vin1* overlaid on *mt1* (transmit carrier) and *vin2* overlaid on *mt2* (transmit carrier with 90 degrees of phase delay). The resulting modulated signals, *vm1* and *vm2*, can be found in the third and fourth plots from the bottom. The summed signal (*rf_out*) can be found in the upper graph. Note the change in phase in the *rf_out* signal every time a new symbol is modulated.

Digital Communication System

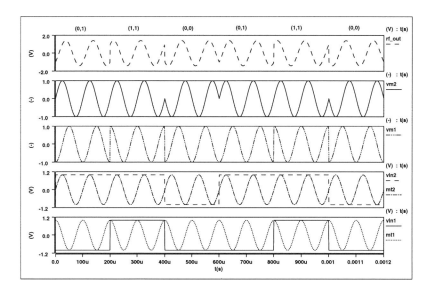

Figure 6.42 Modulator transient simulation results.

In Figure 6.43 the internal demodulator block transient results can be found. The lower graph is the signal *rf_in*. The second and third graphs from the bottom show *vo1* and *vo2*, respectively, which are the individual symbol components after they have been demodulated. The upper two graphs reveal the signals *vout1* and *vout2* which are the filtered symbol components. Note the low-pass filter attenuates the higher harmonics. The harmonics can cause the data slicer to produce bit-errors.

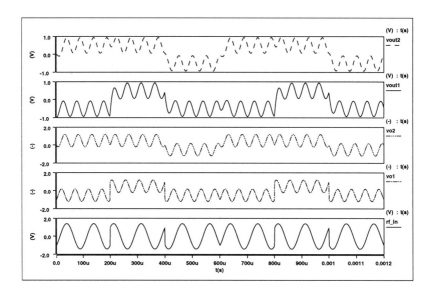

Figure 6.43 Demodulator transient simulation results.

In Figure 6.44 the top-level transient results can be found. The lower two graphs reveal the signals *vout1* and *vout2* which are the inputs to the data slicing comparators. The upper two graphs show an overlay of *vin1* with *vout1* and *vin2* with *vout2*. Note the original bits have been successfully recovered.

Digital Communication System

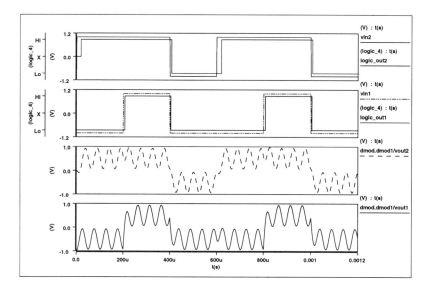

Figure 6.44 Top-level transient results: low-pass filter with a pole at 100 kilo-hertz.

In order to gain a better understanding of how the demodulator low-pass filters affect the overall system, we will modify them to have a DC gain of 120 dB and the signal pole frequency set to 1 mega-hertz.

Again, a DC and transient simulations were performed on the test setup:

```
dc
tr (siglist / /*.*/*,tend 1.2m, tstep.01u).
```

The top-level transient simulation results with altered filters can be found in Figure 6.45. The lower two graphs reveal *vout1* and *vout2;* these signals are the inputs to the data slicing comparators. The upper two graphs show an overlay of *vin1* with *vout1* and *vin2* with *vout2*. Note that the harmonics in *vout1* and *vout2* have greater magnitudes; this causes the comparators to create bit-errors. Each time the comparator's positive input crosses the 0 volt

189

slicing level, the comparator's logic output will change.

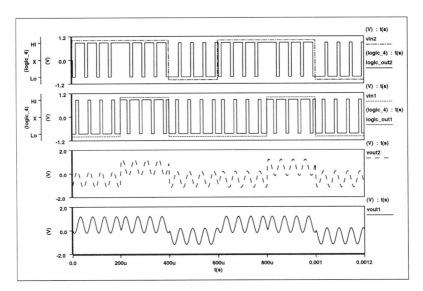

Figure 6.45 Top-level transient results: low-pass filter with a pole at 1 MHz.

This basic example shows that various low-pass filter characteristics in the demodulator can be optimized for the best system performance. An actual QPSK modulator/demodulator system contains much more complexity that involves functions such as forward-error correction, data conversion, up and down frequency conversion, carrier tracking, bit timing, automatic gain control, etc. Despite these complexities, top-level issues can be investigated and isolated through ABM techniques. Issues such as phase noise, third-order intercept point (IP3), inter-modulation, and bit-error rate performance can be analyzed. At the device-level, studying how different mixer implementations (i.e. variations of the Gilbert multiplier) can affect mixing characteristics (e.g. local oscillator leakage) is important. As the low-level issues are understood, they can be incorporated in the top-level analysis; the important concept to remember is to design top-down and verify bottom-up; modeling and simulation can make it all possible!

REFERENCES

[1] Marty Brown, Motorola Semiconductor Products Sector, Tempe, AZ, Personal Communication.

[2] Marty Brown, *Power Supply Cookbook*, Butterworth-Heinemann, Newton, MA, 1994.

[3] John G. Kassakian, Martin F. Schlecht, and George Verghese, *Principles of Power Electronics*, Addison-Wesley Publishing Company, New York, 1991.

[4] Ned Mohan, Tore M. Udeland, and William P. Robbins, *Power Electronics: Converters*, Applications, and Design, John Wiley & Sons, New York, 1989.

[5] Paul Arthur Duran, "Behavioral Modeling of a DC-to-DC Converter", Master's Thesis, Department of Electrical Engineering and Computer Science, Massachusetts Institute of Technology, May 1993.

[6] Analogy Inc., *Power Express^{TM} User's Guide - Release 4.0*, Published with permission from Analogy Inc., Beaverton, OR, 1992.

[7] Trevor Mellard, *Automotive Electronic Systems*, William Heinemann Ltd., London, 1987.

[8] Ken Layne, *Automotive Electronics and Basic Electrical Systems*, John Wiley & Sons, New York, 1990.

[9] John Borwick, *Loudspeaker Headphone Handbook*, Butterworth-Heinemann Ltd.,Boston, 1994.

[10] Mike Donnelly, Analogy Inc., Beaverton, OR., Personal Communication.

[11] Martin Colloms, *High Performance Loudspeakers*, Pentech Press, London, 1985.

[12] Lawrence E. Larson, *RF and Microwave Circuit Design for Wireless Communications*, Artech House, Boston, 1996.

[13] Mischa Schwartz, *Information Transmission, Modulation, and Noise - Fourth Edition*, McGraw-Hill, New York, 1990.

Appendix A

The material presented in this Appendix was published with permission from Analogy, Inc. © 1985-1997. All rights reserved.

DISTRIBUTED POWER SUPPLY EXAMPLE

A.1 Buck Averaged Converter Netlist

```
############################################################
#
# Saber netlist for design bavg
# Created by the Saber Integration Toolkit   4.0-2.8 of
# Analogy, Inc.
# Created on Thu Jan 16 15:48:47 1997.
#
############################################################

############################################################
#
#   Intermediate template comp10
#
############################################################

template comp10 vref:vref vi:vi vo:vo gnd:0 = r4, c1, c2,
r1, r3, r2, vm, vp
#default compensation values
r..rnom r4=100meg
c..c c1=0.1u
c..c c2=0.1u
r..rnom r1=10k
r..rnom r3=10k
r..rnom r2=10k
v..dc vm=-15
v..dc vp=15
```

Appendix A

```
{
r.r4 m:0 p:inm = rnom=r4
c.c1 m:a p:inm = c=c1
c.c2 m:b p:vo = c=c2
r.r1 m:a p:inm = rnom=r1
r.r3 m:vi p:a = rnom=r3
r.r2 m:inm p:b = rnom=r2
v.v2 m:0 p:vee = dc=vm
v.v1 m:0 p:vcc = dc=vp
op1.op1 out:vo inm:inm inp:vref vcc:vcc vee:vee = mod-
el=(a=1meg, rout=10)
}


############################################################
#
#   Instances found in the top level of design bavg
#
############################################################

comp10.comp1 vref:vref vi:vout vo:vc_c gnd:0 = vp=10,\
             vm=0, r2=50, r3=5.38k, \
             r1=119.62k, c2=1479p, c1=618p, r4=390.625k

#compensation values for stable system
#comp10.comp1 vref:vref vi:vout vo:vc_c gnd:0 = vp=10,\
#             vm=0, r2=50k, r3=5k, \
#             r1=10k, c2=1479p, c1=.02u, r4=46.875k

r.rout m:0 p:vout = rnom=.33
v.vin1 m:0 p:vdc_in = tran=(pwl=[0,15])
v.vc1 m:0 p:v_tr = tran=(pwl=[0,0,1u,.7816])
c.c1 m:0 p:vout = esr=.01, c=2.5u
v.ref1 m:0 p:vref = dc=2.5
l.l1 m:vout p:c = l=.53m, r=1m
pwm_buck_cvm.avg1 a:vdc_in c:c cntlm:0 cntlp:vc p:0 = \
                  dutymin=0, dutymax=.98, \
                  rd=60m, cntlmin=0, von=.755, \
                  cntlmax=2.5*.98, rt=55m
```

Forward Averaged Converter Netlist

```
switch_vin.breakpt1 out:vc in1:v_tr in2:v_ac in3:vc_c = \
                    input=use3
v.ac1 m:0 p:v_ac = ac=(mag=-1,phase=0), tran=(off=1)
```

A.2 Forward Averaged Converter Netlist

```
############################################################
#
#  Saber netlist for design favg
#  Created by the Saber Integration Toolkit   4.0-2.8 of
#  Analogy, Inc.
#  Created on Wed Jan 15 17:06:31 1997.
#
############################################################

############################################################
#
#  Intermediate template comp10
#
############################################################

template comp10 vref:vref vi:vi vo:vo gnd:0 = r4, c1, c2, \
               r1, r3, r2, vm, vp
#default compensation values
r..rnom r4=100meg
c..c c1=0.1u
c..c c2=0.1u
r..rnom r1=10k
r..rnom r3=10k
r..rnom r2=10k
v..dc vm=-15
v..dc vp=15

{
r.r4 m:0 p:inm = rnom=r4
c.c1 m:a p:inm = c=c1
c.c2 m:b p:vo = c=c2
r.r1 m:a p:inm = rnom=r1
```

Appendix A

```
r.r3 m:vi p:a = rnom=r3
r.r2 m:inm p:b = rnom=r2
v.v2 m:0 p:vee = dc=vm
v.v1 m:0 p:vcc = dc=vp
op1.op1 out:vo inm:inm inp:vref vcc:vcc vee:vee = \
        model=(a=1meg, rout=10)
}

############################################################
#
#   Instances found in the top level of design favg
#
############################################################

comp10.comp1 vref:vref vi:vout vo:vc_c gnd:0 = vp=10,\
             vm=0, r2=50k, r3=5k, \
             r1=10k, c2=1479p, c1=.02u, r4=7.5k
```

#compensation values for stable system
```
#comp10.comp1 vref:vref vi:vout vo:vc_c gnd:0 = vp=10,\
#             vm=0, r2=50k, r3=5.38k, \
#             r1=119.62k, c2=1479p, c1=618p, r4=62.5k

r.rout m:0 p:vout = rnom=1.5
v.vin1 m:0 p:vdc_in = tran=(pwl=[0,155])
v.vc1 m:0 p:v_tr = tran=(pwl=[0,0,1u,.7927])
c.c1 m:0 p:vout = esr=.01, c=25u
switch_vin.breakpt1 out:vc in1:v_tr in2:v_ac in3:vc_c =\
                    input=use3
pwm_frwd_cvm.avg1 a:vdc_in c:c cntlm:0 cntlp:vc p:0 =\
                  dutymin=0, dutymax=0.98,rd=50m, cntlmin=0,\
                  von=.755, cntlmax=2.5*0.98, n=3
v.ref1 m:0 p:vref = dc=5
v.ac1 m:0 p:v_ac = ac=(mag=-1,phase=0), tran=(off=1)
l.l1 m:vout p:c = l=100u, r=1m
```

A.3 Cascaded Converter Netlist

```
############################################################
#
#   Saber netlist for design dist_tst
#
#   Created by the Saber Integration Toolkit   4.0-2.8 of
#   Analogy, Inc.
#   Created on Thu Jan 16 15:59:10 1997.
#
############################################################

############################################################
#
#   Intermediate template comp10
#
############################################################

template comp10 vref:vref vi:vi vo:vo gnd:0 = r4, c1, c2,\
            r1, r3, r2, vm, vp
#default compensation values
r..rnom r4=100meg
c..c c1=0.1u
c..c c2=0.1u
r..rnom r1=10k
r..rnom r3=10k
r..rnom r2=10k
v..dc vm=-15
v..dc vp=15

{
r.r4 m:0 p:inm = rnom=r4
c.c1 m:a p:inm = c=c1
c.c2 m:b p:vo = c=c2
r.r1 m:a p:inm = rnom=r1
r.r3 m:vi p:a = rnom=r3
```

```
r.r2 m:inm p:b = rnom=r2
v.v2 m:0 p:vee = dc=vm
v.v1 m:0 p:vcc = dc=vp
op1.op1 out:vo inm:inm inp:vref vcc:vcc vee:vee = \
        model=(a=1meg, rout=10)
}

############################################################
#
#   Intermediate template buck
#
############################################################

template buck out:out in:in gnd:0 global_gnd:global_gnd

{
set_l4_1.enb set1:@"n#29"
set_l4_1.init1 set1:@"n#30"
comp_l4.@"comp_l4#27" enbl:@"n#29" out:mod m:vc_c\
        p:@"n#54" = td=1n, hys=1p, \
        enable_init=_1
v.vref1 m:0 p:vref = tran=(pwl=[0,0,.1m,0,1m,2.5])
```

#compensation values for stable system
```
comp10.comp1 vref:vref vi:out vo:vc_c gnd:global_gnd =\
            vp=10, vm=0, r2=50k, r3=5k, \
            r1=10k, c2=1479p, c1=.02u, r4=46.875k
ramposc.ramp1 ct:@"n#54" enable:@"n#30" osc:v_clk\
            rt:@"f#0" gnd:0 = vclow=0, \
            freq=200k, vrt=5, tr=1n, vchigh=2.5,\
            deadt=.0025u
nrlch_l4.rsff1 q:ff_out r:v_clk s:mod qn:@"f#1"
inv_l4.inv1 out:switch in:ff_out
sw_l4.sw1 c:switch m:d p:in = roff=10meg, ron=55m,
tr=100n, tf=100n
d.d1 n:d p:0
c.c1 m:0 p:out = esr=.01, c=2.5u
l.l1 m:out p:d = l=.53m, r=1m
mduty.m1 out:dc in:switch = report_msg="", reject_msg=""
}
```

Cascaded Converter Netlist

```
############################################################
#
#   Intermediate template fwd
#
############################################################

template fwd out:out in:in gnd:0 global_gnd:global_gnd

{
#compenstion values for stable system
comp10.comp1 vref:vref vi:out vo:vc_c gnd:global_gnd =\
             vp=10, vm=0, r2=50k, r3=5.38k, r1=119.62k,\
             c2=1479p, c1=618p, r4=62.5k
set_14_1.enb set1:@"n#35"
set_14_1.init1 set1:@"n#50"
inv_14.inv1 out:switch in:ffout
nrlch_14.rsff1 q:ffout r:v_clk s:mod qn:@"f#0"
sw_14.sw1 c:switch m:a p:in = roff=10meg, ron=1m,\
          tr=100n, tf=100n
sw_14.sw2 c:switch m:0 p:b = roff=10meg,\
          ron=1m, tr=100n, tf=100n
ramposc.ramp1 ct:@"n#54" enable:@"n#50"\
              osc:v_clk rt:@"f#1" gnd:0 = vclow=0, \
              freq=200k, vrt=5, tr=1n, vchigh=2.5,\
              deadt=.0025u
xfrl2.xfmr1 pm:b pp:a sm:0 sp:c = lp=1.8m, k=1, rp=1m,
ls=.2m, rs=1m
c.c1 m:0 p:out = esr=.01, c=25u
v.vref1 m:0 p:vref = tran=(pwl=[0,0,.1m,0,1m,5.0])
d.d1 n:a p:0
d.d2 n:in p:b
d.d3 n:d p:c
d.d4 n:d p:0
l.l1 m:out p:d = l=100u, r=1m
mduty.m1 out:dc in:switch = report_msg="", reject_msg=""
comp_14.@"comp_14#32" enbl:@"n#35" out:mod m:vc_c\
                      p:@"n#54" = td=1n, hys=1p, \
```

```
                    enable_init=_1
}

############################################################
#
#  Instances found in the top level of design dist_tst
#
############################################################

v.v1 m:0 p:vin = dc=155
r.r1 m:0 p:vout = rnom=.33
buck.bck1 out:vout in:dist_v gnd:0 global_gnd:0
fwd.fd1 out:dist_v in:vin gnd:0 global_gnd:0
```

AUTOMOTIVE IGNITION EXAMPLE
A.4 Top-Level Automotive Ignition Netlist

```
############################################################
#
#   Saber netlist for design ./ex_ignit
#   Created by the Netlister Toolkit
#   3.1-1.4 of Analogy,Inc.
#   Created on Sun Feb 9 13:32:06 1996.
#
############################################################

############################################################
#
#  Instances found in the top level of design ./ex_ignit
#
############################################################

short.coil m:coil_short p:coil
torque_w.engine vel1:crankshaft vel2:0 = \
tran=(pulse=(v1=0,v2=500,td=3.6,tr=1,tf=1,pw=10,per=20))
damper_w.damper vel1:crankshaft vel2:0 = d=5
moi_w.inertia vel:crankshaft = j=1
sdr_prsq.start pos:pos = pos-
seq=[(0,1),(1,2),(2,3),(2.2,4),(4,3)]
r.load m:0 p:acc = rnom=2
r.r1 m:plug p:coil_sec = rnom=15k
indx_gen.tdc ip:index vel:crankshaft
ecu_ign.eng out:fire in:ignit index:index = dwell=2m
sparkplug.cyl_1 P1:plug P2:0 = rarc=50k, vsus=300,
vbrk=13k
sw_mtrx3.ign_sw m2:ignit pos:pos m3:sole p1:batt\
```

```
        p2:0 p3:0 m1:acc\ =
        mtrx=[(),(s11=on),(s11=on,s12=on),\
        (s11=off,s12=on,s13=on)], tdbrk=10m, \
        rfunc=discont, ron=1m, tdmk=30m
v.v1 m:0 p:batt_term = dc=13.5
r.batt m:batt_term p:batt = rnom=50m
r.r2 m:0 p:plug_esr = rnom=20k
d1n3012.d1 n:coil_short p:0
q2n6043.q1 coil_short fire 0
dc_srs.starter s2:0 a1:start wrm:crankshaft = laa=50m,\
        j=.01, ra=0.005, lff=10m, laf=10m
d1n3879.d2 n:start p:0
c.c1 m:plug_esr p:plug = c=50p
xfr.xfmr_1 pm:coil pp:batt sm:coil_sec sp:0 = lp=5m,\
        k=0.9, rp=2, ls=50, rs=100
rly_1pno.rly_1 drvp:sole m:start p:batt drvm:0 =\
                roff=1meg, vdrop=5, lcoil=20m, \
                vpull=6, ron=5m
```

A.5 Electronic Control Unit Model Code

```
template ecu_ign in index out = dwell, adv_delay
electrical in, out
state nu index
number dwell = 2m
number adv_delay = 0

{
r.rin in out = 470
r.rdiv out 0 = 470
sw_1pnc.sw1 sw_pos out 0 = ron = 1m, roff = 1meg

when (dc_init) {
schedule_event(time,sw_pos,1)
}

when (event_on(index)) {
if (index == 1) {
schedule_event(time + adv_delay,sw_pos,2)
schedule_event(time + adv_delay + dwell,sw_pos,1)
```

```
}
}
}
```

A.6 Position Sensor Model Code

```
template indx_gen vel ip = thetainit
rotational_vel vel
state nu ip
number thetainit = 0 # Initial angle (in radians).
{
var rad theta
val nu ctheta
val nu velocityp1
val nu stepsizeval
state nu before, after

<consts.sin

when (dc_init) {
schedule_event(time,ip,0)
}

when (threshold(ctheta,0.5,before,after)) {
if(after == 1) {
schedule_event(time,ip,1)
schedule_event(time+1m,ip,0)
}
}

values{
ctheta = cos(theta)
velocityp1 = w_radps(vel) + 1
stepsizeval = 0.2/velocityp1
step_size = stepsizeval
#if (w_radps(vel) ~= 0) step_size = 0.2/w_radps(vel)
#else step_size = 1
}

equations{
```

```
theta: d_by_dt(theta) = (1-dc_domain)*w_radps(vel) +
(dc_domain)*(theta - thetainit)
}
}
```

A.7 Spark Plug Model Code

```
template sparkplug p1 p2 = vbrk, vsus, rarc
electrical p1, p2
number vbrk = 10k,# Breakdown Voltage.
vsus = 500,# Minimum arc sustaining voltage.
rarc = 10k# Resistance of plug during spark.
{
number  tarc_on = 1u,# Arc turn-on and turn-off transition
times.
tarc_off = 1u

number rhigh = 100meg# Nominal "no-arc" resistance.

state time tbreak  # Time of voltage breakdown.
state time theal# Time arc drops out.

var i i# Current, only needed to prevent a
# divide by zero (if used v/r method) during
# the iterations.
val v v # Input voltage difference.
val r r# Instantaneous resistance value.

state nu rstate# State defines what is happening to
# resistance r.
#1 ---- r = rhigh
#2 ---- r = transition to low (rarc)
#3 ---- r = rarc
#4 ---- r = transition to high (rhigh)

state nu before, after# Threshold polarity

state r rlast# Value of r at beginning of transition.
```

Spark Plug Model Code

```
when(dc_init) {
rstate = 1
}

when(threshold(abs(v),vbrk,before,after) &
(rstate == 1 | rstate == 4)) {
if(after == 1) {
tbreak = time
rlast = r
schedule_event(time,rstate,2)
schedule_event(time + tarc_on,rstate,3)
schedule_next_time(time)
schedule_next_time(time + tarc_on)
}
}

when(threshold(abs(v),vsus,before,after) &
(rstate == 3 | rstate == 2)) {
if(after == -1) {
theal = time
rlast = r
schedule_event(time,rstate,4)
schedule_event(time + tarc_off,rstate,1)
schedule_next_time(time)
schedule_next_time(time + tarc_off)
}
}

values {
v = v(p1) - v(p2)
if (rstate == 1) r = rhigh
else if (rstate == 3) r = rarc
# Transition to low (rarc)
else if (rstate == 2) {
r = rlast - (rlast-rarc)*(time - tbreak)/tarc_on
}
# Transition to high (rhigh)
else if (rstate == 4) {
r = rlast + (rhigh-rlast)*(time - theal)/tarc_off
}
```

204

Appendix A

```
}
equations {
i(p1->p2) += i
i: v = i*r
}
}
```

AUDIO TEST SYSTEM EXAMPLE

A.8 Top-Level Audio Test System Netlist

```
############################################################
#
#   Saber netlist for design ex_audio
#   Created by the Saber Integration Toolkit
#   4.0-2.8 of Analogy, Inc.
#   Created on Tue Mar  4 16:09:34 1997.
#
############################################################

############################################################
#
#   Intermediate template osc
#
############################################################

template osc out:out

{
buf_l4.b2 out:rc2 in:ndout = tphl=10.3u, tplh=0.05u
buf_l4.b1 out:rc1 in:i1out = tphl=10.3u, tplh=0.05u
buf_l4.b3 out:rc3 in:out = tphl=2.1u, tplh=2.1u
prbit_l4.start out:start = bits=[(0,_0),(1p,_1)]
inv_l4.i2 out:out in:rc2 = tplh=10n
inv_l4.i1 out:i1out in:rc3 = tplh=10n
nand2_l4.nd1 out:ndout in1:rc1 in2:start = tplh=10n
}

############################################################
#
#   Intermediate template dsp
#
```

Top-Level Audio Test System Netlist

```
############################################################

template dsp dsp_out:dsp_out d0:d0 d1:d1 d2:d2 d3:d3 \
        d4:d4 d5:d5 d6:d6 d7:d7 eoc:eoc gnd:0

{
zlti.zlti1 smp:smp zout:zlti_out zin:zin = a=0.036706,\
        den=[1,-1.864734,0.869309], \
        num=[1,-1.981254,0.986114],\
        model=(td=0,max=10,min=-10,bits=16)
clk2smp.c2s1 smp:smp clk:eoc = delay=100n
z2a.z2a1 aout:dsp_out zin:zout gnd:0 = tt=100n
zlcmb.add1 zout:zout zin1:zlti_out zin2:dly_out =\
        b=0.3, a=0.7, \
        model=(td=0,max=5,min=-5,bits=8)
zdelay.dly1 smp:smp zout:dly_out zin:zlti_out = k=500
b2z.b2z1 d10:@"f#0" d11:@"f#1" d12:@"f#2" d13:@"f#3"\
        d14:@"f#4" zout:zin \
        d15:@"f#5" dsign:@"f#6" d0:d0 d1:d1 d2:d2 \
        d3:d3 d4:d4 d5:d5 clk:eoc d6:d6 d7:d7 \
        d8:@"f#7" d9:@"f#8" = bits=8, code=nob
}

############################################################
#
#   Intermediate template n8div
#
############################################################

template n8div div_out:div_out clk:clk

{
inv_l4.inv1 out:clkn in:clk = tplh=5n
set_l4_1.set1 set1:set1
dff_l4.dff1 d:qn1 q:q1 r:set1 s:set1 qn:qn1 clk:clkn =\
        qinit=_0, tp=10n
dff_l4.dff2 d:qn2 q:q2 r:set1 s:set1 qn:qn2 clk:q1 =\
        qinit=_0, tp=10n
dff_l4.dff3 d:qn3 q:div_out r:set1 s:set1 qn:qn3 clk:q2 =\
        qinit=_0, tp=10n
```

Appendix A

}

```
############################################################
#
#   Intermediate template csp
#
############################################################

template csp vin:vin vout:vout gnd:0

{
r.rout m:vout p:v_oc = rnom=50
r.rin m:0 p:vin = rnom=50
var2elec.v2e1 in:filt_out m:0 p:v_oc = k=2
src.src1 out:pulse = \
         tran=(pulse=(v1=0,v2=1,tr=1u,tf=1u,\
         td=10m,pw=50m,per=200m))
limit.lim1 out:lim_out in:pulsetone = lim=5, k=1
lag.lag1 out:filt_out in:lim_out = w=6283, k=1
mult.mult1 out:pulsetone in1:cin in2:pulse
elec2var.e2c1 out:cin m:0 p:vin = k=1
}

############################################################
#
#     Intermediate template d2a
#
############################################################

template d2a out:out d0:d0 d1:d1 d2:d2 d3:d3 d4:d4 d5:d5
d6:d6 d7:d7 gnd:0

{
r.r3 m:opin p:v3 = rnom=160k
r.r2 m:opin p:v2 = rnom=320k
r.r1 m:opin p:v1 = rnom=640k
```

Top-Level Audio Test System Netlist

```
r.r0 m:opin p:v0 = rnom=1280k
r.r7 m:opin p:v7 = rnom=10k
r.r6 m:opin p:v6 = rnom=20k
r.r5 m:opin p:v5 = rnom=40k
r.r4 m:opin p:v4 = rnom=80k
r.r_fb m:out p:opin = rnom=5k
sw_l4.sw_3 c:d3 m:v3 p:vnref = roff=10meg, ron=1m
sw_l4.sw_2 c:d2 m:v2 p:vnref = roff=10meg, ron=1m
sw_l4.sw_1 c:d1 m:v1 p:vnref = roff=10meg, ron=1m
sw_l4.sw_0 c:d0 m:v0 p:vnref = roff=10meg, ron=1m
sw_l4.sw_7 c:d7 m:v7 p:vnref = roff=10meg, ron=1m
sw_l4.sw_6 c:d6 m:v6 p:vnref = roff=10meg, ron=1m
sw_l4.sw_5 c:d5 m:v5 p:vnref = roff=10meg, ron=1m
sw_l4.sw_4 c:d4 m:v4 p:vnref = roff=10meg, ron=1m
v.vcc m:0 p:vcc = dc=5
v.vee m:0 p:vee = dc=-5
v.nref m:0 p:vnref = dc=-15
op1.op_1 out:out inm:opin inp:vee vcc:vcc vee:vee
}

############################################################
#
#   Intermediate template adc
#
############################################################

template adc vin:vin start:start eoc_out:eoc_out d0:d0\
            d1:d1 d2:d2 d3:d3 \
            d4:d4 clk:clk d5:d5 d6:d6 d7:d7 gnd:0


{
r.rin m:0 p:vin = rnom=50
set_l4_0.set0 set0:set_0
buf_l4.buf1 out:eoc_out in:eoc = tplh=100n
shft8_l4.sh1 load:eoc q0:d0 q1:d1 q2:d2 q3:d3 q4:d4 \
        q5:d5 q6:d6 q7:d7 ld0:q0 \
        ld1:q1 ld2:q2 ld3:q3 ld4:q4 ld5:q5 ld6:q6 ld7:q7
clk:set_0 sdin:set_0 \
          clr:set_0 clkinh:set_0 = tp=10n
d2a.d2a1 out:vout d0:q0 d1:q1 d2:q2 d3:q3 d4:q4 \
```

```
            d5:q5 d6:q6 d7:q7 gnd:0
sar_bhv.sar1 q0:q0 q1:q1 q2:q2 start:start q3:q3 \
             q4:q4 q5:q5 q6:q6 q7:q7 \
             data:data clk:clk eoc:eoc = delay=100n
comp_l4.cmp1 enbl:@"f#0" out:data m:vout p:vin = \
             td=10n, enable_init=_1
}

###########################################################
#
#   Intermediate template rlc
#
###########################################################

template rlc vin:vin vout:vout gnd:0   = rpot
r..rnom rpot=1k

{
r.r1 m:mid_l p:vin = rnom=100
r.r2 m:0 p:vout = rnom=rpot
l.l1 m:vout p:mid_l = l=25m
c.c1 m:0 p:vout = c=1u
}

###########################################################
#
#   Intermediate template pwramp
#
###########################################################

template pwramp out_p:out_p vin:vin vcc:vcc\
                out_m:out_m gnd:0

{
r.r_leak m:0 p:out_m = rnom=1meg
r.re m:0 p:ve = rnom=3.3, ratings=(pdmax_ja=5)
r.r2 m:0 p:vb = rnom=220
r.r1 m:vb p:vcc = rnom=1k
c.cin m:vin p:vb = c=100u
xfr.x1 pm:vc pp:vcc sm:out_m sp:out_p = lp=400m,\
```

Top-Level Audio Test System Netlist

```
        k=0.995, rp=100m, \
        ratings=(pdmax_ja=30, vpmax=40, vsmax=10,\
        ipmax=5,ismax=20), ls=100m, rs=50m
q_3p.q1 b:vb c:vc e:ve = ratings=\
        (pdmax_ja=10,icmax=5,vcemax=50), \
        model=(type=_n,bf=150,is=3e-12,\
        rb=2,rc=100m,re=50m,cje=1n,cjc=500p,ikf=3,br=5)
}

##########################################################
#
#  Intermediate template lspkr
#
##########################################################

template lspkr dia_pos:dia_pos in_m:in_m in_p:in_p gnd:0

{
spring_nl.susp pos1:dia_pos pos2:0 = k3=95meg,\
               k1=2k, delta0=0
voice_coil.vc1 m:in_m p:in_p pos1:dia_pos pos2:0 =\
               b=300m, l=1m, len=13, r=8
winddrag.air pos:dia_pos = w=0.1, d=0.1
mass.mdia pos:dia_pos = m=0.016


##########################################################
#
#  Instances found in the top level of design ex_audio
#
##########################################################

osc.osc1 out:clk
dsp.dsp1 dsp_out:dsp_out d0:d0 d1:d1 d2:d2 d3:d3 d4:d4\
         d5:d5 d6:d6 d7:d7 eoc:eoc gnd:0
n8div.n8div1 div_out:clk_div8 clk:clk
r.rin m:vsrc p:vsrc_oc = rnom=50
v.vsrc m:0 p:vsrc_oc = \
       ac=(mag=1,phase=0),\
       tran=(sin=(va=7,f=100,vo=0))
```

210

Appendix A

```
csp.csp1 vin:vsrc vout:vtest gnd:0
adc.adc1 vin:vtest start:clk_div8 eoc_out:eoc \
         d0:d0 d1:d1 d2:d2 d3:d3 d4:d4 \
         clk:clk d5:d5 d6:d6 d7:d7 gnd:0
rlc.rlc1 vin:dsp_out vout:filt_out gnd:0 = rpot=1k
v.v_pwr m:0 p:v_pwr = dc=30
pwramp.pa1 out_p:pa_out vin:filt_out vcc:v_pwr\
           out_m:pa_ref gnd:0
lspkr.lspkr1 dia_pos:diaphragm in_m:pa_ref \
             in_p:pa_out gnd:0
```

A.9 Loudspeaker Subsystem Test Netlist

```
############################################################
#
#   Saber netlist for design ex_lspkr
#   Created by the Saber Integration Toolkit
#   4.0-2.8 of Analogy, Inc.
#   Created on Thu Mar  6 23:05:03 1997.
#
############################################################


############################################################
#
#   Instances found in the top level of design ex_lspkr
#
############################################################

v.vin m:0 p:vin = ac=(mag=1,phase=0), \
      tran=(pulse=(v1=0,v2=4k,tr=0.1u,\
      tf=0.1u,td=0,pw=0.25m))
spring_nl.susp pos1:diaphragm pos2:0 = k3=95meg,\
               k1=2k, delta0=0
voice_coil.vc1 m:0 p:v_coil pos1:diaphragm pos2:0 =\
               b=300m, l=1m, len=13, r=8
r.rsrc m:v_coil p:vin = rnom=8
winddrag.air pos:diaphragm = w=0.1, d=0.1
```

DSP Subsystem Test Netlist

```
mass.mdia pos:diaphragm = m=0.016
```

A.10 DSP Subsystem Test Netlist

```
############################################################
#
#   Saber netlist for design ex_dsp
#
#   Created by the Saber Integration Toolkit   4.0-2.8 of
#   Analogy, Inc.
#   Created on Thu Mar  6 22:39:50 1997.
#
############################################################


############################################################
#
#   Instances found in the top level of design ex_dsp
#
############################################################

set_l4.fix_d7 set:d7 = level=_1
clock_l4.clk1 clock:eoc = freq=5k
prbit_l4.pr_d3 out:d0_6 =
bits=[(0,_0),(100u,_1),(500u,_0)]
zlti.zlti1 smp:smp zout:zlti_out zin:zin = a=0.036706,\
          den=[1,-1.864734,0.869309], \
          num=[1,-1.981254,0.986114], mod-
el=(td=0,max=10,min=-10,bits=16)
clk2smp.c2s1 smp:smp clk:eoc = delay=100n
z2a.z2a1 aout:dsp_out zin:zout gnd:0 = tt=100n
zlcmb.add1 zout:zout zin1:zlti_out zin2:dly_out = \
          b=0.3, a=0.7,model=(td=0,max=5,min=-5,bits=8)
zdelay.dly1 smp:smp zout:dly_out zin:zlti_out = k=500
b2z.b2z1 d10:@"f#0" d11:@"f#1" d12:@"f#2" d13:@"f#3"\
         d14:@"f#4" zout:zin d15:@"f#5" \
         dsign:@"f#6" d0:d0_6 d1:d0_6 \
```

```
            d2:d0_6 d3:d0_6 d4:d0_6 d5:d0_6\
            clk:eoc d6:d0_6 d7:d7 d8:@"f#7" \
            d9:@"f#8" = bits=8, code=nob
```

A.11 Voice Coil MAST Model Code

```
###########################################################
#
#   This template created by Analogy, Inc.
#   Copyright 1995 Analogy, Inc.
#
###########################################################

element template voice_coil p m pos1 pos2 = b,len,r,l

electrical p,m              # Electrical pins.

translational_pos pos1,     # Position of coil.
                  pos2      # Reference position.


number b = 300m,    # Flux density across the air-ga (Teslas)
len = 15,           # Length of the wire in the coil (meters)
r = 8,              # Coil "DC" electrical resistance (Ohms)
l = 1m              # Coil inductance (H)

{
var i i                 # Current in the coil (A)
var vel_mpsvel          # Velocity of coil, (meters/sec)
val pos_mposn           # Differential position (coil to
ref, meters)
val frc_Nforce          # Electro-magnetic force, N
val vv_bemf             # Back EMF generated voltage
values {
# Define differential position
posn = pos_m(pos1) - pos_m(pos2)
# Electro-magnetic force generated by current in coil
force = b*len*i
# Calculate back EMF voltage due to coil motion
```

```
v_bemf = b*len*vel
}
equations {
i(p->m) += i
frc_N(pos1->pos2) += force
i: v(p)-v(m) = r*i + d_by_dt(l*i) + v_bemf
vel: vel = d_by_dt(posn)
}
}
```

A.12 Wind-Drag MAST Model Code

```
############################################################
#
#           This template created by Analogy, Inc.
#       for exclusive use with the Saber(TM) simulator.
#               Copyright 1991 Analogy, Inc.
#
############################################################

############################################################
#
# Template winddrag.sin, a winddrag and viscous damper
# model with   mechanical translation
# position connection.
#
# This model represents a winddrag and viscous damper. Its
# behavior is such that a translational velocity on the
# input connection produces a damping force related
# to that velocity:
#
# force = -1*d*velocity - w*velocity*|velocity|
#
#
# Note the "-" sign in the force equation. This is
# consistant with the convention that a positive force
# causes the position(pos) to increase. In this case,
# the damping force tends to resist a positive
# translational velocity.
```

Appendix A

```
# The connection pos is a "translational" pin, with
# translational position as the across variable
# and force as the through variable.
# These variables are specified in MKS units, where the
# position is in meters and force is in Newtons.
# However, you can select other units for parameter
# specifications and for viewing the variables
# in the simulation results. This is described below.
#
# The viscous damping constant d can have any
# value (except "inf"). The default value is d = 0.
# The winddrag constant w can also have any value.
# However, you must assign a value, as there is no
# default. The model interprets the value specified
# for d as being in some particular units. These units
# depend on the selection of the "units" parameter,
# as follows;
#
#if units =   d is interpreted as: w is interpreted as:
#----------   --------------------- -------------------
#    mks       (Newtons)/(meter/sec) N/(m/sec)**2
#    cgs       (dynes)/(cm/sec)     dynes/(cm/sec)**2
#    fps       (pounds)/(ft/sec)    pounds/(ft/sec)**2
#    ios       (ounces)/(in/sec)    ounces/(in/sec)**2
#
# NOTE: units = mks is the default.
#
# Following simulation, the internal variables
# representing position, velocity and force can be
# viewed by calling out the appropriate group
# name in the signal list. Group names and associated
# position units are;
#
# group name:position:velocity:force:
# ------------------------------------------
# mks_values  meters   m/sec     Newtons
# cgs_values  centimeters cm/sec dynes
# fps_values  feet     ft/sec    pounds
# ios_values  inches   in/sec    ounces
#
```

Wind-Drag MAST Model Code

```
############################################################


element template winddrag pos = d, w, units

translational_pos pos    # Position connection.

number d = 0,            # Translational damping constant.
w                        # Translational winddrag constant.

mech_def..units units = mks # Selects "units"
                            # assumed for arguments.

{

number d_eff             # Effective (mks) equivalent of d.
number w_eff             # Effective (mks) equivalent of w.

val pos_mposn_mks        # Position (meters).
val pos_cposn_cgs        # Position (cm).
val pos_fposn_fps        # Position (ft).
val pos_iposn_ios        # Position (in).

var vel_mpsvel_mks       # Velocity (meters/sec).
val vel_cpsvel_cgs       # Velocity (cm/sec).
val vel_fpsvel_fps       # Velocity (ft/sec).
val vel_ipsvel_ios       # Velocity (in/sec).

val frc_Nfrc_mks         # Total damping force (Newtons).
val frc_dfrc_cgs         # Total damping force (dynes).
val frc_pfrc_fps         # Total damping force (pounds).
val frc_ofrc_ios         # Total damping force (ounces).

group {posn_mks,vel_mks,frc_mks} mks_values
group {posn_cgs,vel_cgs,frc_cgs} cgs_values
group {posn_fps,vel_fps,frc_fps} fps_values
group {posn_ios,vel_ios,frc_ios} ios_values

parameters {
```

Appendix A

```
if (d == inf | d == undef) {
saber_message("TMPL_S_ILL_VALUE",instance(),"d")
}

if (w == inf | w == undef) {
saber_message("TMPL_S_ILL_VALUE",instance(),"w")
}

if (units == mks) {
d_eff = d                  # No conversion necessary.
w_eff = w                  # No conversion necessary.
}

else if (units == cgs) {
d_eff = d * 1e-3           # Convert dyne/(cm/sec)
                           # to N/(m/sec).
w_eff = w * 1e-1           # Convert dyne/(cm/sec)**2
                           # to N/(m/sec)**2.
}

else if (units == fps) {
d_eff = d * 14.594         # Convert lb/(ft/sec) to N/(m/sec).
w_eff = w * 47.879         # Convert lb/(ft/sec)**2
                           # to N/(m/sec)**2.
}

else if (units == ios) {
d_eff = d * 10.945         # Convert oz/(in/sec) to N/(m/sec).
w_eff = w * 430.91         # Convert oz/(in/sec)**2
                           # to N/(m/sec)**2.
}
}

values {
posn_mks = pos_m(pos)   # Position (meters).

frc_mks = -1*d_eff*vel_mks - w_eff*vel_mks*abs(vel_mks)

#### Convert variables to alternate units for display.
```

```
posn_cgs = posn_mks * 100      # Convert m to cm.
posn_fps = posn_mks * 3.2808   # Convert m to ft.
posn_ios = posn_fps * 12       # Convert ft to in.

vel_cgs = vel_mks * 100        # Convert m/s to cm/s.
vel_fps = vel_mks * 3.2808     # Convert m/s to ft/s.
vel_ios = vel_fps * 12         # Convert ft/s to in/s.

frc_cgs = frc_mks * 1e5        # Convert N to dynes.
frc_fps = frc_mks * 0.22481    # Convert N to pounds.
frc_ios = frc_fps * 16         # Convert pounds to ounces.
}

equations {
frc_N(pos) += frc_mks
vel_mks: vel_mks = d_by_dt(posn_mks)
       }
}
```

A.13 Successive Approximation Register MAST Model Code

```
############################################################
#
# This is a behavioral model of an 8-bit successive
# approximation register (SAR). It will perform an
# A-to-D conversion (when used with a d2a and a
# comparator) in 8 clock cycles. The parameter "delay"
# is the time from "clock" or "start"  to new digital
# output. It must be > 0 but much less than the clock
# period. EOC occurs 2xdelay after last clock rising
# edge, and resets delay/2 after start rising edge.
#
############################################################


template sar_bhv data start clk q7 q6 q5 q4 q3 q2 q1 q0\
              eoc = delay
state logic_4 data,start,clk,q7,q6,q5,q4,q3,q2,q1,q0,eoc
```

Appendix A

```
number delay = 10n      # Defines delay of output and eoc
pulse
{
state nu clk_cnt        # Clock counter
state logic_4 clk_enbl  # Clock enable

when(dc_init) {
clk_enbl = l4_0
}

when(event_on(start)) {
if (start == l4_1) {
clk_cnt = 0
clk_enbl = l4_1
schedule_event(time+delay/2,eoc,l4_0)
schedule_event(time+delay,q7,l4_0)
schedule_event(time+delay,q6,l4_1)
schedule_event(time+delay,q5,l4_1)
schedule_event(time+delay,q4,l4_1)
schedule_event(time+delay,q3,l4_1)
schedule_event(time+delay,q2,l4_1)
schedule_event(time+delay,q1,l4_1)
schedule_event(time+delay,q0,l4_1)
}
}

when(event_on(clk)) {
if (clk == l4_1 & clk_enbl == l4_1) {
clk_cnt = clk_cnt + 1
if (clk_cnt == 1) {
schedule_event(time+delay,q7,data)
schedule_event(time+delay,q6,l4_0)
}
else if (clk_cnt == 2) {
schedule_event(time+delay,q6,data)
schedule_event(time+delay,q5,l4_0)
}
else if (clk_cnt == 3) {
schedule_event(time+delay,q5,data)
schedule_event(time+delay,q4,l4_0)
```

```
    }
    else if (clk_cnt == 4) {
    schedule_event(time+delay,q4,data)
    schedule_event(time+delay,q3,14_0)
    }
    else if (clk_cnt == 5) {
    schedule_event(time+delay,q3,data)
    schedule_event(time+delay,q2,14_0)
    }
    else if (clk_cnt == 6) {
    schedule_event(time+delay,q2,data)
    schedule_event(time+delay,q1,14_0)
    }
    else if (clk_cnt == 7) {
    schedule_event(time+delay,q1,data)
    schedule_event(time+delay,q0,14_0)
    }
    else if (clk_cnt == 8) {
    schedule_event(time+delay,q0,data)
    schedule_event(time+2*delay,eoc,14_1)
    clk_enbl = 14_0
    }
  }
 }
}
```

A.14 Digital to Z-Domain Converter Model Code

```
###########################################################
#
# This template converts a convention, digital clock signal
# into a z-domain type sampling signal. On the rising edge
# of the dig. clock ([0,x,z] to 14_1), the z output
# changes state. The output looks like the frequency/2, but
# its sampling effect is at the original clock frequency.
#
###########################################################
```

Appendix A

```
template clk2smp clk smp = delay
state logic_4 clk
state nu smp
number delay = 0 # Delay of smp after clk edge
{
state logic_4 clk_old# Last clock value
state nu smp_old# Last smp value (0 or 1)

when(dc_init) {
smp_old = 0
}

when(event_on(clk)) {
if (clk == l4_1 & clk_old ~= l4_1) {
if (smp_old == 0) {
schedule_event(time+delay,smp,1)
smp_old = 1
}
else if (smp_old == 1) {
schedule_event(time+delay,smp,0)
smp_old = 0
}
}
clk_old = clk
}
}
```

A.15 Nonlinear Spring MAST Model Code

```
############################################################
#
# Template spring_nl.sin, a non-linear spring model with
# translational position connections. This model
# represents a spring. Its behavior is such that
# a relative displacement (pos1 - pos2) between the input
# connections produces a restraining force:
#
# force = -1*k1*[(pos1-pos2)-delta0] - k3*[(pos1-pos2)-
```

Nonlinear Spring MAST Model Code

```
#              delta0]**3
#
###########################################################

element template spring_nl pos1 pos2 = k1, k3, delta0

translational_pos pos1,    # Position connection 1.
                  pos2     # Position connection 2.

number k1,         # Linear spring coefficient.
       k3 = 0,     # Third power spring coefficient.
delta0 = 0         # Equilibrium position of spring (where
force = 0).

{

val pos_mposn_mks   # Extension (compression) in meters
val frc_Nforce_mks  # Restraining force, N.

values {
posn_mks = pos_m(pos1)-pos_m(pos2)
force_mks = -1*k1*(posn_mks-delta0)-k3*(posn_mks-
delta0)**3
}

equations {
frc_N(pos1->pos2) += force_mks
         }
}
```

Appendix A

DIGITAL COMMUNICATIONS SYSTEM

A.16 Top-level Digital Communications System Netlist

```
###########################################################
#
#   Saber netlist for design tel_test
#   Created by the Saber Integration Toolkit 4.0-2.8 of
#   Analogy, Inc.
#   Created on Sun Mar 23 23:49:18 1997.
#
###########################################################

###########################################################
#written template hins
#written template hins1
###########################################################

###########################################################
#
#   Intermediate template dmod
#
###########################################################

template dmod vout1:vout1 vout2:vout2 rf_in:rf_in gnd:0

{
elec2var.ecin out:rf_inc m:0 p:rf_in = k=1
elec2var.ec5 out:mr1c m:0 p:mr1
elec2var.ec6 out:mr2c m:0 p:mr2
var2elec.ec1 in:vc_out1 m:0 p:vout1 = k=1
var2elec.ec2 in:vc_out2 m:0 p:vout2 = k=1
vsine.vm1 m:0 p:mr1 = ph=90, ampl=1, f=10k
vsine.vm2 m:0 p:mr2 = ampl=1, f=10k
mult.m3 out:vo1 in1:rf_inc in2:mr1c
mult.m4 out:vo2 in1:rf_inc in2:mr2c
```

Top-level Digital Communications System Netlist

```
hins.@"ratiosc#47" out:vc_out1 in:vo1
hins1.@"ratiosc#48" out:vc_out2 in:vo2
}

############################################################
#
#   Intermediate template mod
#
############################################################

template mod vin1:vin1 vin2:vin2 rf_out:rf_out gnd:0
{
vsine.vm1 m:0 p:mt1 = ph=90, ampl=1, f=10k
vsine.vm2 m:0 p:mt2 = ampl=1, f=10k
var2elec.ec1 in:vt m:0 p:rf_out = k=1
elec2var.ec4 out:mt2c m:0 p:mt2
elec2var.ec3 out:mt1c m:0 p:mt1
elec2var.ec2 out:vin2c m:0 p:vin2
elec2var.ec1 out:vin1c m:0 p:vin1
sum.s1 out:vt in1:vm1 in2:vm2
mult.m2 out:vm2 in1:vin2c in2:mt2c
mult.m1 out:vm1 in1:vin1c in2:mt1c
}


############################################################
#
#   Instances found in the top level of design tel_test
#
############################################################

dmod.dmod1 vout1:vout2 vout2:vout1 rf_in:rf gnd:0
mod.mod1 vin1:vin1 vin2:vin2 rf_out:rf gnd:0
v.vin2 m:0 p:vin2 = \
        tran=(pwl=[0,-1,1u,1,399u,1,400u,-1,599u,\
              -1,600u,1,999u,1,1000u,-1])
v.@"v_pwl#99" m:0 p:vin1 = \
        tran=(pwl=[0,-1,199u,-1,200u,1,399u,1,400u,\
              -1,799u,-1,800u,1,999u,1,1000u,-1])
```

Digital Communication System

```
comp_14.@"comp_14#108"  enbl:@"n#116"  out:logic_out1 m:0\
                        p:vout2 = td=1n, hys=.2
comp_14.@"comp_14#109"  enbl:@"n#117"  out:logic_out2 m:0\
                        p:vout1 = td=1n, hys=.2
set_14_1.@"set_14_1#115"  set1:@"n#117"
set_14_1.@"set_14_1#114"  set1:@"n#116"
```

Index

A
ABDC 16
ABM methodology 53
abstraction 23
AHDL 16
AHDL simulator 7
algorithmic modeling 35
analog behavioral macromodeling 34
analog behavioral modeling 33
analog simulation 5
analog-control algorithm 11
analog-to-digital converter 120, 168
analog-to-digital interface model 127
AND gate 106
audio test system 164
automotive ignition system 154

B
battery 157
bottom-up design 44
bottom-up verification 164, 190
buck converter 139, 141, 148

C
C++ 35
capacitor 76
clock generator 167
coil driver 158
conceptual-level modeling 27
crossover filter 170
current-controlled current source 65

D
DC current source 63
DC motor 131
DC voltage source 63
device-level modeling 26
digital communication system 181
digital HDLs 13
digital signal processor 169
digital-control algorithm 11
digital-to-analog interface model 129
distributed power supply 138
D-latch 109
Dolphin Integration 16
dynamic specification 36

E
Eldo 20
electronic control unit (ECU) 157
event-driven simulator 9
exponential sinusoid 66

F
forward converter 140, 146, 149
frequency-to-voltage converter 102
functional-level modeling 26

H
hardware description language (HDL) 12
HDL-A 20
hierarchical design 47
high-level description languages (HLDs) 27

I
IC modeling 30
ideal diode 77
ideal transformer 85
ideal transistor 79
IEEE 1076 14
ignition coil 157
ignition switch 157

inductor 77

L
lock-step algorithm 11
loudspeaker 171

M
MAST 21
MAST mini-tutorial 59
MAST template 60
MATHCAD 27
Matlab 27
mixed-mode HDL simulator 10
model accuracy 28
model development 55
model documentation 56
model precision 29
model specification 54
model verification 56
modeling continuum 25
motor crankshaft 158
multi-level simulation 51
multiplexer 108

N
native mixed-signal HDL simulator 12
netlist 12
Newton-Raphson 7

O
Open Verilog International 15

P
peak detector 89
position sensor 158
power amplifier 171

pulse-width modulator 116

Q
QPSK demodulator 185
QPSK modulator 184
QPSK transmission 181

R
release control 56
resistor 75
rollback algorithm 11

S
sample-and-hold 92
Schmitt trigger 96
SMASH 16
solenoid/relay 157
spark plug 158
SpectreHDL 17
SPICE 6, 8, 27
SPICE device modeling 30
SPICE macromodeling 32
SPICE Simulator 6
SPW 27
starter motor 157
system-level modeling 27

T
test-tone generator 167
top-down design 45, 163, 180, 190

V
Verilog 15
Verilog-A 17
Verilog-AMS 19
VHDL 13
VHDL-AMS 19

Index

VHSIC 13
virtual test 38
voltage arithmetic 69
voltage comparator 113
voltage-controlled voltage source 64
voltage differentiation 72
voltage integration 74
voltage-to-frequency converter 99

W
winddrag 172

ABOUT THE AUTHOR

Paul A. Duran obtained his BSEE and MSEE from the Massachusetts Institute of Technology, Cambridge, MA, in 1992 and 1993, respectively. He also obtained an MBA from the Arizona State University, Tempe, AZ, in 1996. Mr. Duran has held positions as an Automated Design and Systems Engineer at the Motorola Semiconductor Products Sector specializing in modeling and simulation of IC systems. He has system experience with switching power supplies, servo-control systems, storage systems, automotive control systems, and data conversion systems. Mr. Duran has also worked for Rockwell Semiconductor Systems where he has held positions in Technical Marketing supporting broadband communication products. Mr. Duran is currently working for the Applied Micro Circuits Corporation (AMCC) as an Applications Engineer specializing in high-speed data communication systems for HDTV, SAN, LAN, MAN, and WAN products.